Fuels and Petrochemicals Division 2018

Core Programming Area at the 2018 AIChE Annual Meeting

Pittsburgh, Pennsylvania, USA
28 October - 2 November 2018

ISBN: 978-1-5108-7619-4

Printed from e-media with permission by:

Curran Associates, Inc.
57 Morehouse Lane
Red Hook, NY 12571

Some format issues inherent in the e-media version may also appear in this print version.

Printed by Curran Associates, Inc. (2019)

For permission requests, please contact AIChE
at the address below.

AIChE
120 Wall Street, FL 23
New York, NY 10005-4020

Phone: (800) 242-4363
Fax: (203) 775-5177

www.aiche.org

Additional copies of this publication are available from:

Curran Associates, Inc.
57 Morehouse Lane
Red Hook, NY 12571 USA
Phone: 845-758-0400
Fax: 845-758-2633
Email: curran@proceedings.com
Web: www.proceedings.com

TABLE OF CONTENTS

Some abstracts in this book refer to images that were not made available by the publisher

540773 2018 Outlook for Energy: A View to 2040

Abstract:

The Outlook for Energy is ExxonMobil's global view of energy demand and supply through 2040. It also highlights the dual challenge of providing affordable energy to support prosperity while reducing environmental impacts – including the risks of climate change. We use the Outlook to help inform our long-term business strategies and investment plans.
The Outlook anticipates significant changes through 2040 across the world to boost living standards, reshape the use of energy, broaden access to abundant energy supplies, and accelerate decarbonization of the world's energy system to address the risks of climate change.

- Global energy needs will rise about 25 percent over the period to 2040
- Electricity demand will nearly double in non-OECD countries
- Worldwide electricity from solar and wind will increase about 400 percent
- Natural gas will expand its role, led by growth for electricity generation
- Growth in oil demand will be driven by commercial transportation and chemical needs
- Efficiency gains and growing use of less-carbon intensive energy sources will contribute to a nearly 45 percent decline in the carbon intensity of global GDP

Energy is fundamental to modern life, and as the world's population approaches 9 billion people in 2040, we are challenged to improve living standards everywhere. Understanding the factors that drive the world's energy needs – and likely choices to meet those needs – is the mission of the Outlook. By sharing the Outlook with the public, we hope to broaden that understanding. Energy matters to everyone, and we all play a role in shaping its future.

541543 Energy Decarbonisation Scenarios

Abstract:

With the energy sector being associated with 2/3 of the annual greenhouse gas emissions, decarbonizing the energy is a key step to reduce their level. The National Determined Contributions (NDC's) announced so far by the participating countries do not meet the objectives of the Paris-climate agreement, as far as reaching a peak in emissions in the near future or decreasing to a level that is compatible with the stated long-term targets.

Using different scenarios for energy demand and supply, options for energy sector decarbonisation are evaluated. Meeting the 2-degree target requires a major transformation in the power sector, efficiency improvements – specially in the buildings and in the industry, and a revolution in transport towards fuel and modal shifts. With a significant increase in renewable energy, system integration and optimization will also be key in reducing the cost of the energy transition. Meeting more ambitious climate goals ("well below 2 degrees") will push further the technological innovation and implementation requirements throughout the entire energy value chain. A scenario aiming at zero net-emissions by 2060 is presented, along with a comparison to the 2-degree case.

542625 Fundamental Research Needs to Advance Energy Technologies

Abstract:

The U. S. Department of Energy (DOE), Office of Science, Office of Basic Energy Sciences (BES) supports fundamental research in chemical and materials sciences to provide the foundations for new energy technologies and to support DOE missions in energy, environment, and national security. This presentation will discuss opportunities for fundamental research to impact DOE's energy mission "to catalyze the timely, material, and efficient transformation of the nation's energy system and secure U.S. leadership in energy technologies" with a focus on the way we generate, store and use energy nationally. I will provide an overview of BES strategic planning over the past decade that identified priority research directions for advancing energy applications, highlight key scientific advances in these areas, and discuss some future opportunities for modern science, particularly chemical sciences, to accelerate the transformation of the U. S. energy portfolio.

Methaforming Produces Gasoline from Naphtha and Methanol at 1/3 the Current Cost

By Stephen Sims, Iosif Lishchiner, Olga Malova
New Gas Technologies – Synthesis LLC, USA/Russia

Corresponding author: Stephen Sims, s.sims@ngts.us

Summary

Methaforming removes sulfur and converts naphtha and methanol into a high-octane, low-benzene gasoline blendstock with yields and octane comparable to isomerization + CCR reforming. Methaforming uses a proprietary zeolite catalyst in a process flow similar to hydrotreating.

This one-step process replaces naphtha desulfurization, reforming, isomerization and benzene removal thereby reducing capital and operating costs to 1/3 of the current best alternative.

An existing idle hydrotreater or reformer can be converted into a commercial Methaformer at very low cost.

Main Applications of Methaforming:

1. Building a Methaformer instead of a CCR reformer: worth $220 million net present value (NPV) for a 20 K BPD (860 K tpa) unit.

2. Converting semi-regen reformer: worth $50 million/year at a retrofit cost of $20 million for a 20 K BPD (860 K tpa) unit.

3. Instead of isom: worth $10 million/year with $20 million lower capex for a 10 K BPD (380 K tpa) unit.

4. Upgrade raffinate from aromatics extraction: upgrade 2 K BPD (88 K tpa) worth $18 million/year at a capex of $30 million.

Methaforming Process and Economics

Figure 1 Methaforming: One-Step Conversion

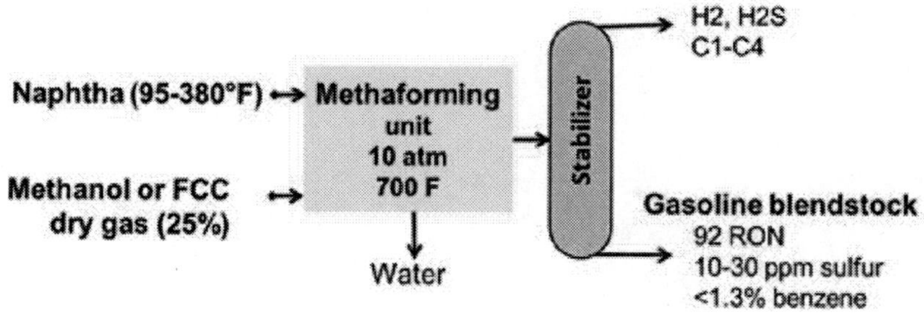

The 1/3 cost can be accomplished because Methaforming is a one-step process. Naphtha and methanol go through a unit similar to a hydrotreater at modest pressure and temperature. The products are like reformate: a high-octane gasoline blendstock with relatively low sulfur, and importantly, <1.3% benzene. Just like a reformer it produces some light ends, and releases hydrogen. And by desulfurizing, there is H_2S in the overhead. The inexpensive proprietary zeolite catalyst does not contain precious metals.
The capital and operating costs of Methaforming are comparable to those of a hydrotreater. The process flow in which methanol is used instead of hydrogen, looks like that of a hydrotreater without a recycle compressor. The yields are comparable to a CCR reformer except that most of the benzene is converted to toluene and half the methanol becomes water.

Figure 2 Methaforming vs Reforming: Simpler and Cheaper

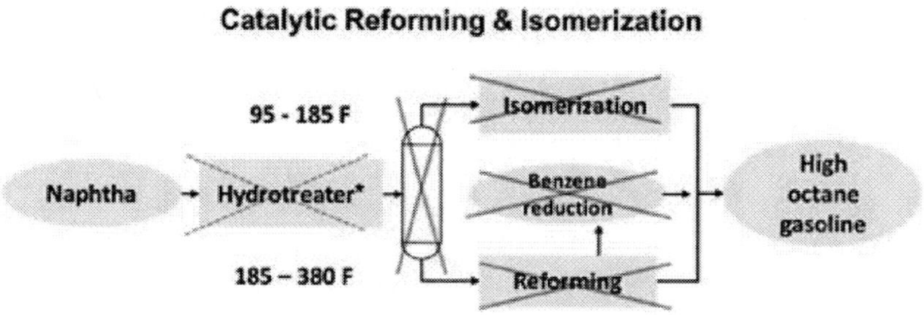

Methaforming costs are so much lower because a Methaformer replaces four units. The Methaformer charges naphtha, typically full range naphtha, and reduces sulfur by 90%. It replaces the reformer, and because of low benzene production, avoids the need for any benzene reduction steps. Methaforming also effectively processes light naphtha, thereby eliminating the isomerization unit. In this way capital and operating costs are reduced to about 1/3.

Figure 3 Economics of Methaformer vs Traditional Reforming Suite (HDS+CCR+isom)

New 20 K BPD unit (860 K tpa)	Methaforming	Alternative	Δ Methaforming-Alternative
Yields, $MM/yr	206	202	+ 4
OpEx, $MM/yr	7	21	- 14
CapEx, $MM	50	156	- 106
Total NPV, $MM	1 300	1 080	**+ 220**

The economics for a grassroots 20 K BPD Methaformer are shown in the first column of Figure 3. The first row is for the yields including fuel gas. The next rows are for other operating expense and finally the estimated capex for inside battery limits (ISBL) only. The last row of the Methaforming column shows the total 20 year NPV for a Methaformer is $1.3 billion. The next column shows the same values for the alternative of HDS + CCR + isom. The last column shows the difference between Methaforming and the alternative process.

The Methaformer is slightly better in yields at $4 million/year. Opex is $14 million/year lower because less equipment is needed. The capex is $106 million less for ISBL. This gives a NPV difference of $220 million.

Figure 3 shows that by using Methaforming yields are comparable to the CCR reformer while the operating costs and capex are much lower.

Because few grassroots reformers are being built, the largest market for Methaformers is in replacing existing semi-regen reformers.

Figure 4 Economics for Upgrade Semi-regen Reformer to Methaformer

Existing 20 K BPD unit (860 K tpa)	Methaforming	Alternative	Δ Methaforming-Alternative
Yields, $MM/yr	206	149	+ 57
OpEx, $MM/yr	7	14	- 7
CapEx, $MM	20	-	- 20
Total NPV, $MM	1 330	920	**+ 410**

Figure 4 uses the same format as Figure 3 to show the economics. Focusing on the Delta column, Methaformer yields generate over $50 million/year higher than for semi-regen reforming. It costs $20 million to convert to a Methaformer. As a result the Methaformer shows over $400 million NPV. Because a Methaformer is similar to a hydrotreater, the unit that is converted is the naphtha hydrotreater in front of the reformer.

Methaforming is Flexible

Methaforming is very flexible so it generates several additional economic opportunities.

FCC olefins may be used to replace the methanol in a Methaformer. While refinery propylene is usually recovered and upgraded, many refineries leave the ethylene ($C2^{=}$) in the fuel gas stream. There is a 500 $/mt uplift for ethylene in a Methaformer thereby increasing uplift on naphtha by over $11/B ($100/mt).

Methaforming gives the opportunity to lower sulfur while increasing octane for the natural gasoline from a natural gas liquids (NGL) plant and for cracked naphthas.

Figure 5 Economics for Methaforming Light Virgin (LVN) Naphtha with $C2^{=}$ vs Isomerization

New 10 K bpd unit (380 K tpa)	Methaforming	Alternative (isom)	Δ Methaforming-Alternative
Yields, $MM/yr	120	110	+ 10
OpEx, $MM/yr	4	6	- 2
CapEx, $MM	30	50	- 20
Total NPV, $MM	860	750	**+ 110**

There is keen interest in the US and Western Europe for processing LVN because the higher anticipated octane requirements drive consideration of isomerization with expensive recycle. Processing LVN with $C2^=$ shows over $35/bbl uplift ($330/mt). In a side-by-side comparison, Methaforming with $C2^=$ shows $10 million/yr better yields giving a $110 million better NPV.

Figure 6 Economics for Grassroots Methaformer to Process Raffinate and FCC Dry Gas

New 2 K BPD unit (88 K tpa)	Methaforming	Alternative (direct blending)	Δ Methaforming-Alternative
Yields, $million/yr	89	60	+29
OpEx, $million/yr	2	0	+2
CapEx, $million/yr	17	0	+17
Total NPV, $million	160	n/a	**+160**

Refiners with low octane streams blended to gasoline (e.g. raffinate from aromatics extraction) are interested in Methaforming. For a 2 K BPD (88k tpa) stream of raffinate, Methaforming with the added benefit of $C2^=$, gives a 1/2-year payback and $160 million NPV for this small unit from uplift of $42/B ($350/mt).

Process Development and Testing

Figure 7 Pilot Plants

The Methaforming catalyst and process were developed, and the operating parameters optimized, in 3 pilot plants with over 7000 hours of processing. Our latest and largest pilot plant started up in 2015 and has a 2 liter reactor and capacity of a 38 l/d (0.23 BPD). Based on the extensive pilot plant testing, we can reliably predict the results for any naphtha.

A major Russian refiner who plans to build an 11 K BPD (470 K tpa) Methaformer for start-up in 2020 has funded a new, larger and more sophisticated pilot plant to confirm the design. This pilot plant, to be located in Moscow at the High Temperature Research Institute, will start up in October 2018.

Its configuration will be a 3-stage 9 L reactor with a capacity of up to 2 BPD (90 tpa) designed for continuous operation and testing of the products. It will include enhanced features to enable co-feeds of methanol and/or other alcohols of varying concentrations, and olefin gases such as in FCC dry gas. Two additional features will address stricter gasoline specifications: on-purpose hydrotreating capability to increase desulfurization from 90 to 99% and benzene recycle for <1% in the product Methaformate.

The NGTS intellectual property includes 4 Russian patents with PCT coverage that are entering the international phase, plus 2 recent PCT applications. The catalyst is produced by an experienced manufacturer of zeolite catalysts and is protected by trade secrets.

Figure 8 First Commercial Methaformer

The first commercial Methaformer is a 150 BPD (6 k tpa) unit will processing a variety of naphtha feedstocks. With a projected uplift of >$23/B ($200/ton), it will validate the predicted yields and scale-up factors.

Because of Methaforming's similarity to the hydrotreating process, there is a low-cost and high return opportunity by converting an existing idle hydrotreater or reformer. The basic cost for this might be as low as $5 million, mostly for the methanol handling facilities. Because most refiners have some low octane naphtha being blended to gasoline, those with an idle process unit have a very low-cost opportunity for upgrading that stream.

Figure 9 Diagram of Skid Mounted Methaformer

Many potential customers are interested in quick delivery, low-cost skid mounted units. We have completed the detailed mechanical design for 2 versions. The contractor will build these in Russia at a fixed price of $900 K for a 150 BPD (6K tpa) unit and $2 million for a 460 BPD (20 K tpa) unit. These can be delivered within 6 months and will include all of the equipment necessary inside battery limit (ISBL). The outside battery limit is dependent on existing customer facilities. Much of this may already be available or can be available on an incremental cost basis.
Our first customer in the CIS will soon order a 400 BPD (20 K tpa) unit.

Conclusion

Methaforming upgrades low octane streams by $15-35/B ($130-330/mt) thereby producing gasoline from naphtha and methanol at 1/3 the cost of current technology for new units. It can generate over $50 million/year of yield improvements by converting an existing semi-regen reformer.

531271 Oxidative Desulphurisation Using Vortex Based Hydrodynamic Cavitation with Water Dispersed in Organic Phase

Abstract:

Oxidative Desulphurisation using a Vortex Based Hydrodynamic Cavitation

Abatement of sulphur from diesel is a growing concern, displayed by implementation of legislation in many countries around the globe, the EU, China and USA have all enforced a sulphur concentration limit of 10 – 15 ppm for diesel [1]. To refine diesel to these standards, hydrodesulphurisation (HDS) the industry standard, is being operated at extreme conditions, (100 – 200 atm, 320 – 400 ^{o}C) for several cycles, significantly driving up cost as consumption of H_2 and MoS_2 catalyst is augmented [2]. Even with these conditions HDS struggles to reduce the sulphur concentration under 30 ppm, considering this coupled with the cost of an additional HDS unit, (approx. $50 – 80 million) an alternative process is widely sought after [3].

Thus, various other desulphurisation technologies are currently being researched and developed: Adsorptive Desulphurisation (ADS), Bio-Desulphurisation (BDS), Extractive Desulphurisation (EDS) and Oxidative Desulphurisation (ODS) are the four of the main processes under discussion and test. Of these, ODS has been identified as the more effective process from a performance and economical viewpoint, due to mild operating conditions; along with usage of common refinery equipment and complimentary chemistry to HDS [3].

ODS traditionally uses an oxidant/catalyst often coupled with an extractant to aid in separation of Sulphur containing compounds from fuels. Ultrasound Assisted Oxidative Desulphurisation (UAOD) has received attention from several research groups, consisting of exposing the reaction mixture to an ultrasonic wave which enhances the formation of OH· due to inception of cavities via *'Acoustic Cavitation' (AC).* AC is known to intensify a process by inducing micro-bubbles or cavities which develop extreme localised conditions (> 10,000K & 1000 atm) following bubble growth and collapse, forming the aforementioned radicals which dissociate water molecules existing in the liquid medium [4].

AC has successfully enhanced ODS on a laboratory scale, however, there are serious issues with scale-up and associated cost limitations [5]. Another form of cavitation, *'Hydrodynamic Cavitation' (HC),* demonstrates high potential to induce cavitation in a more energy efficient manner. HC exhibits extreme potential due to its effectiveness **without the use of external oxidant or catalyst.**

Research carried out by Suryawanshi et al. [6] have shown that HC can effectively remove ≥ 99% of refractory sulphur compounds from both model oil and diesel when low organic volume fractions (2.5%) are instigated. Their work also investigated important parameters such as identification of cavitation inception point, influence of pressure drop, initial sulphur concentration, volume fraction of organic phase and effect of model-oil species/sulphur species amongst others. The concept exhibits high potential since it does not use any external oxidant or catalyst. This work utilises only a small volume fraction of organic phase suspended in aqueous phase. Considering the large volumes of transportation fuels, this approach may demand excessively large reactors. In this research, we have developed the concept of using hydrodynamic cavitation for oxidative desulphurisation further by operating it as aqueous phase dispersed in organic phase. This approach has a potential to significantly improve the overall productivity.

The process of cavitation in a two-phase system with aqueous phase suspended in an organic phase will be investigated systematically. Model oil composed of a heavy hydrocarbon, such as dodecane, will make-up the organic phase. Model sulphur compounds thiophene, dibenzothiophene (DBT) and 4,6 – dimethyldibenzothiophene (4,6 – DMDBT) are considered as the sulphur compounds in this work as they have been identified as key S containing compounds in transportation fuels.

Experiments will be carried out for investigating:

- The influence of pressure drop, volume fraction of aqueous phase and initial sulphur concentration.
- Effect of suspended aqueous phase on cavitation inception, yield and collapse intensity.
- Overall desulphurisation efficiency for the above parameters.

Benchmarking of results will be carried out and compared to a batch system treated by AC, utilising an acoustic horn. The AC process will be carried out utilising a fluid volume of 200ml, a frequency set to 20kHZ, amplitude of 80% and reaction time of 1 hour, whilst maintaining a temperature beneath 50°C. All results will be analysed by an Agilent 7890A GC with FPD and HP – 1 (*30m x 0.32mm x 0.3mm)* column.

Cavity dynamics models will be adapted for the liquid-liquid dispersed systems. The cavity dynamics models will be used to interpret the experimental results. This will greatly facilitate further development and optimisation of oxidative desulphurisation using hydrodynamic cavitation.

A main feature of the research is utilisation of a 3D-printed vortex-based cavitation device. This design shields the walls from collapsing cavities and offers significantly better performance than conventional hydrodynamic cavitation devices (Surywanshi et al. [7]). This design incorporated into this research aims to continue development of a highly effective oxidative desulphurisation process by bridging the gaps in fundamental understanding of cavitation in multiphase flow and its influence on ODS. The results presented here will provide a sound basis for further work on development of optimal and industrially sustainable process oxidative desulphurisation process.

References

1. https://www.dieselforum.org. (2016). Diesel Sulfur Limits Worldwide and the Need for ULSD. [online] Available at: https://www.dieselforum.org/policyinsider/diesel-sulfur-limits-worldwide-and-the-need-for-ulsd.

2. Babich, I. (2003). Science and technology of novel processes for deep desulfurization of oil refinery streams: a review . Fuel, 82(6), pp.607-631.

3. Ibrahim, M., Hayyan, M., Hashim, M. and Hayyan, A. (2017). The role of ionic liquids in desulfurization of fuels: A review. Renewable and Sustainable Energy Reviews, 76, pp.1534-1549.

4. Ja'fari, M., Ebrahimi, S. and Khosravi-Nikou, M. (2018). Ultrasound-assisted oxidative desulfurization and denitrogenation of liquid hydrocarbon fuels: A critical review. Ultrasonics Sonochemistry, 40, pp.955-968.

5. Anderson, K.; Atkins, M.P.; Borges, P.; Chan, Z.P; Rafeen, M.S.; Sebran, N.H; van der Pool, E.; Vleeming, J.H. Economic analysis of ultrasound-assisted oxidative desulfurization, Energy Source, Part B: Economics, Planning and Policy, 2017, 12 (4), 305–311.

6. *Suryawanshi, N., Bhandari, V., Sorokhaibam, L. and Ranade, V. (2016). A Non-catalytic Deep Desulphurization Process using Hydrodynamic Cavitation. Scientific Reports, 6(1).*

7. *Developing techno-economically sustainable methodologies for deep desulfurization using hydrodynamic cavitation./ Suryawanshi, Nalinee B.; Bhandari, Vinay M.; Sorokhaibam, Laxmi Gayatri; Ranade, Vivek V.* **In:** *Fuel, Vol. 210, 15.12.2017, p. 482-490.*

533207 Low Pressure Hydrodeoxygenation of Liquid Phase Pyrolysis Oil and Refinery Intermediates

Abstract:

Various concepts have been developed to cover the increasing energy demand in transportation. Conventional combustion engines are not expected to be replaced by alternatives in the near future. This makes the production of fuels and substitute fuels out of biogenic feedstock indispensable. A suitable pathway for biofuel production is direct biomass liquefaction via pyrolysis.

In liquid phase pyrolysis[1], biomass is processed by a liquid heat carrier. In the bioCRACK process[2,3], crude oil refinery intermediates such as vacuum gas oil are used as heat carrier. The bioCRACK process combines the pyrolysis of biomass with the cracking of the heat carrier. Non-polar biomass fragments are directly transferred into the non-polar heat carrier oil. The biogenic carbon load of the heat carrier oil is processed in the refinery without further treatment. Polar biomass fragments are transferred into the liquid phase pyrolysis oil (LPP oil). Due to the high water content, high corrosivity and other negative properties, LPP oil needs an intensive upgrade prior to usage as fuel for combustion engines.[4] To achieve these standards, hydrodeoxygenation (HDO) is proposed.

LPP oil was further upgraded in a lab scale hydrodeoxygenation (HDO) reactor. To reduce investment costs for industrial application, a crude oil refinery compatible process is required. Therefore, low pressure co-hydrodeoxygenation of liquid phase pyrolysis oil and heavy gas oil was investigated.

HDO was carried out in a plug flow reactor from Parr Instrument Company with inline-sulfided metal-oxide catalyst in two steps. Direct Co-Processing of LPP oil with refinery intermediates led to plugging. To avoid plugging, low temperature HDO prior to Co-processing for dewatering and hydrophobising was suggested. Both steps were performed at 80 bar and at liquid hourly space velocity (LHSV) of 1 h^{-1}. In the first step (HDO 1), LPP oil was hydrodeoxygenated at 300°C, in the second step (HDO 2), the product of HDO 1 was mixed with heavy gas oil (10 wt.% HDO1/ 90 wt.% heavy gas oil) and hydrodeoxygenated at 400°C. After the first step, dewatering and hydrophobation of LPP oil was achieved, making the product of HDO 1 miscible with heavy gas oil. In the second step, LPP oil was finally upgraded to a product with fuel quality (Table 1).

Table 1. Elemental composition and water content of educts and products compared to diesel

HDO	Water content [wt.%]	C [wt.%]	H [wt.%]	Rest = O [wt.%]	N [wt.%]
LPP oil	58.9	20.8	9.4	69.5	<1
Heavy gas oil	0.0	85.4	14.6	0.0	<1
HDO 1	1.95	76.3	11.6	11.7	<1
HDO 2	0.01	85.3	14.7	0.0	<1
Diesel	<0.02	86.3	13.7	0.0	<1

1 N. Schwaiger, D. C. Elliott, J. Ritzberger, H. Wang, P. Pucher and M. Siebenhofer, *Green Chem.*, 2015, **17**, 2487â€"2494.

2 K. Treusch, J. Ritzberger, N. Schwaiger, P. Pucher and M. Siebenhofer, *R.Soc.open sci.*, 2017, **4**.

3 J. Ritzberger, P. Pucher and N. Schwaiger, 2014, **39**, 1189â€"1194.

4 Q. Zhang, J. Chang, T. Wang and Y. Xu, *Energy Convers. Manag.*, 2007, **48**, 87â€"92.

539384 Plausible Pathway to Meet IMO 2020 Global Sulfur Cap

Abstract:

Following the International Maritime Organizationâ€™s (IMO) decision to decrease the global limit for sulfur content of marine fuel oil from 3.5 wt% to 0.5 wt% effective January 1 2020, there has been a lot of discussion about how ship owners may comply. Several options have been proposed to meet this new limit for the shipping industry; utilizing on-board scrubbers with the existing high sulfur fuel oils, switching to LNG fuels which are practically sulfur free, or switching to low sulfur fuels by blending. For refiners; another option may be further processing of the high sulfur crude residue through challenging technology such as resid hydroprocessing. These options all have advantages and disadvantages, some require expensive retrofit of the current ships, some come with higher risks due to limited experience, however, the biggest constraint is the time to enforcement. Ultimately refiners as well as the transportation industries are looking for the most economical pathway for immediate solution.

Historically, marine fuel oil, also known as bunker fuel, is the atmospheric tower bottoms (ATB), which is the feedstock for the vacuum tower and is seldom available for fuel oil blending; thus the bunker fuel oil would typically be a blend of any of the following streams- low sulfur ATB, FCC slurry oil, hydrotreated gas oils, coker liquid, visbroken residue, deasphalted oil (DAO) etc. However, other studies indicate that, despite the fact that pricing of the blend components is an issue; the availability of the blend components that could meet the bunker fuel recipe in all geographic areas of the world is a greater problem. As for example, most of the low-sulfur crudes in the world are available outside the United States. Besides that, not all refineries have all the units or plants to produce the blend materials.

Typically, refiners fractionate the crude feed before hydroprocessing the different fractions. The heaviest fractions such as the ATB and vacuum resid (VR) that typically go into the bunker fuel recipe are the most difficult to hydroprocess. In this paper we explored an unorthodox option as a possible solution to meeting the demand of reduced sulfur bunker fuel market targeting the global limit of IMO - crude oil pretreatment through hydrotreating. The most common hydrotreater unit in a refinery may be used to produce low-sulfur blend material directly from whole crudes. This may eliminate the major capital cost of installing a new unit.

At the University of Tulsa Research Campus, we have a state-of-the-art large scale (4 to 4 ½ L) continuous hydroprocessing facility that could be operated up to 540°C and 27.6 MPa pressure. The present paper will discuss the unusual option of hydrotreating and subsequent fractionation of a whole crude oil with focus on the extent of Hydrodesulfurization (HDS) and Hydrodenitrification (HDN) of its ATB and/or VGO fractions. The experimental system was operated at 0.5 h^{-1} LHSV, 15.2 MPa and H_2/oil ratio of 1200 L/L and 3 temperatures (335°C, 348°C and 357°C) using commercially available Nickel-Molybdenum on alumina catalyst. The feed and products were analyzed and bulk conversion of each contaminant as well as the density change is presented. The extent of HSD in the resulting products make them good candidates for blending materials to meet the target <0.5 wt% sulfur. The practicability of this option and other options will also be discussed.

529813 Modification of FCC Slurry Oil Used for Producing High-Grade Paving Asphalt

Abstract:

Fluid catalytic cracking (FCC) slurry oil is regarded as an ideal soft component for asphalt blending process due to its low cost and high content of aromatics. However, the bad ageing resistance limits its application to only low-grade paving asphalt production. A new process has been proposed where propanal is used as the highly active slurry oil modifier with small molecular structure and low molecular weight. With the aid of the modifier, the active groups in the slurry oil are deactivated and the ageing resistance of FCC slurry oil is improved.

The FCC slurry oil has been modified with modifier in a batch reactor according to the defined mass ratio and the modified slurry oil has been characterized by FT-IR, NMR and kinematic viscosity. The main conditions in the slurry oil modification process have been investigated, including the reaction temperature, pressure, time, modifier contents and other factors. The results demonstrated the occurrence of the electrophilic substitution reaction during the modification. When reaction temperature was 180â„ƒ, reaction pressure was 2.0 MPa, the reaction time was 5h, and the modifier contents was 3.0(wt%), the kinematic viscosity of modified slurry decreased from 129.21 mm^2/s to 64.51 mm^2/s, and the penetration ratio of the modified slurry blended asphalt increased from 53.9 to 65.3. The asphalt using modified slurry oil as soft component can be achieved to meet standard requirements of 70# paving asphalt. FCC slurry oil modification technology provides a new method for both SDA and asphalt process economically.

529992 Spherical Polymer Brushes Bearing Pyrrolidone Groups As Novel Nickel Remover for Crude Oil

Abstract:

Today, fossil fuels are still the main sources of the energy demand. However, metals in heavy oils create significant detrimental impact on hydrocracking and cracking processes. Especially, Nickel(Ni) poison the catalyst by accumulating on its surface and lead to the reduction of light oil yield. Therefore, it becomes a hot issue to be solved urgently in crude oil processing. Currently demetallizers are mainly organic acids and their salts. Due to their strong acidity, adding these demetallizers will cause the corrosion of equipment. Nowadays polymer brush materials with a core-shell structure show good performance in the fields of biomedicine, homogeneous catalysis, heavy metals adsorption and other fields.

In our previous research(Fuel,2018,226:47-53), polymer brushes bearing imidazole groups has good nickel efficiency. By further discussion and research, it is found that pyrrolidone(VP) groups have many excellent physical and chemical properties, adsorption and good thermal stability. In this work, N-vinyl pyrrolidone(VP) is grafting onto the SiO_2 using â€œgrafting toâ€ method to build spherical polymer brushes with strong chelating ability on nickel (PVP@SiO_2). PVP@SiO_2 is applied to remove trace nickels in crude oil with the help of electric demulsification. The results show that when the concentration of PVP@SiO_2 is 5000ppm, the ratio of water to oil is 1/10, the shear strength of emulsifier is 8 kr/min, the shear time was 1.5 min, the electric desalting temperature is 120â„ƒ, the high voltage is 1000V/cm, low voltage is 500V/cm and settling time is 60 min, the maximum crude oil de-nickel rate reaches about 64.4%. In addition, the regeneration of polymer brushes are achieved by a rapid and simple process using diluted HCl. As well as, the removal efficiency of Ni remains 58.9% in 6 consecutive cycles. VP, PVP and PVP@SiO_2 show different removal capabilities and the sequence is VP < PVP < PVP@SiO_2. Ultraviolet-visible spectrophotometry and mass spectrometry were employed to characterize the molecular structure of nickel porphyrins, and in combination with the properties of the spherical polymer brush material and the Donnan effect, a possible nickel removal mechanism by PVP@SiO_2 is proposed.

535683 Investigation of Effects of Subcritical Water As Fracturing Fluid on Hydraulic Fracturing and Fracture Permeability of Shale

Abstract:

Hydraulic fracturing is a promising technique to produce unconventional oil and gas. However, with the current hydraulic fracturing practice, only 10-30 % of the hydrocarbon is recoverable from unconventional shale according to the U.S. Energy Information Administration (EIA). Low porosity (<5 %) and low permeability (2-50 nD) of Utica and Marcellus shale are the primary reasons for the low hydrocarbon recovery. Hydrocarbon recovery can be enhanced significantly by connecting the pores and creating new fracture branches which will result in enhanced porosity and enhanced permeability respectively.

Hydraulic fracturing uses water (>90%) as fracturing fluid at temperatures in the range of 50-95 °C. Physical and chemical properties of hot compressed water (T=100-374 °C and corresponding saturation pressure) and supercritical water (T>374 °C and P>22.4 MPa) are very different than water below 100 °C. Thermodynamic properties of water (e.g., density, ionic product, and dielectric constant) changes significantly with the increase of water temperature. As a result, sub-and supercritical water behaves like a strong nonpolar solvent. Supercritical water also becomes highly reactive and corrosive depending on the density and temperature. Therefore, using subcritical water as fracturing fluid is likely to enhance the hydrocarbon recovery from shale reservoirs by enhancing the shale porosity and permeability.

In previous study, shale samples were treated in subcritical water environment which was named as hydrothermal deformation (HTD). With HTD, shale porosity was enhanced significantly. For example, HTD at 300°C for six hours of residence time shale porosity increased up to four times than raw shale. A positive correlation of shale porosity with HTD reaction time and temperature was observed. Evolution of numerous micro and mesopores were also observed.

In this study, HTD was applied in Utica shale samples to observe the changes in shale fracture. It was observed that fracture width of shale was increased significantly with HTD. Besides, additional fractures were also observed. Along with the HTD of shale, subcritical water was also injected into the pre-drilled Utica shale core plug in a core holder with certain confining pressure. Confining pressure was allowed to vary within a range of 2500-3500 psig. Consistent changes in the shale morphology were also found from the SEM-EDS analysis. Digital microscope and micro computerized tomography scanning (CT-scan) images were used to observe the changes in fracture have occurred across the shale core.

531920 Drag Model Development and 3-Phase Simulation of Methane Production from a Gas Hydrate Reservoir

Abstract:

Drag Model Development and 3-Phase Simulation of Methane Production from a Gas Hydrate Reservoir

Deniz Hinz, Hamid Arastoopour and Javad Abbasian

Department of Chemical and Biological Engineering

Wanger Institute for Sustainable Energy Research (WISER)

Illinois Institute of Technology, Chicago, IL

ABSTRACT

Global estimates of resource-grade natural gas from hydrates range from 10,000 trillion cubic feet (tcf) to greater than 100,000 tcf. When compared with the approximately 2200 tcf of U.S. technically recoverable natural gas reserves as of 2012, one can grasp the incredible scale and conceivable global impact of hydrates. In recent years, the 570 tcf of technically recoverable shale gas resources has provided the U.S. considerable economic stimulus and affordable means of decreasing coal consumption. This means the potential impact of hydrate resources will result in very significant economic growth. As the world undergoes the transition into a sustainable energy society, natural gas is the cleanest fossil energy option.

The theoretical potential of hydrates is immense, but production testing and research remain lacking, which has led to the development of numerous hydrate production numerical simulators for consolidated porous medium hydrate reservoirs; however considerable evidence exists suggesting unconsolidated flow behavior. Production data from the JOGMEC Mallik 2L-38 well in Alaska reported extensive sand production, leading to repeated pump failure in the 2007 testing period and the installation of sand screens prior to the successful 2008 testing period.

Pore-scale inhomogeneities intrinsic to the reservoir are magnified during the hydrate dissociation and sand production processes, causing homogeneous permeability models such as the Kozeny-Carman equation to deviate significantly from experimental results. Therefore, a three-phase flow model which includes momentum and continuity equations for each phase and constitutive models for the drag force between gas and liquid phases with the hydrate-bearing sediment were developed. Our drag force models quantify the pore-scale inhomogeneity and simulate the evolution of high-permeability regions in the reservoir. Our numerical simulation prediction of the methane production agreed with the production data from the JOGMEC Mallik 2L-38 well in Alaska.

BADAK LNG PROCESS TRANSFORMATION: CHALLENGES IN HANDLING LEANER FEED GAS

Robby Sukma Dharmawan, PT Badak NGL, Bontang, Indonesia

Introduction

Background

Badak LNG plant, located in Bontang, East Kalimantan, Indonesia, has been excellently operated for more than four decades to process rich gas from Mahakam Block Field. Declining feed gas supply had become the major issue to its operational sustainability after reaching maximum production of 22.5 MTPA LNG in 2001 as shown in **Figure 1**. In 2016, the number of running trains was only four out of eight. It was predicted that there would be no more than 3 running trains beyond 2018.

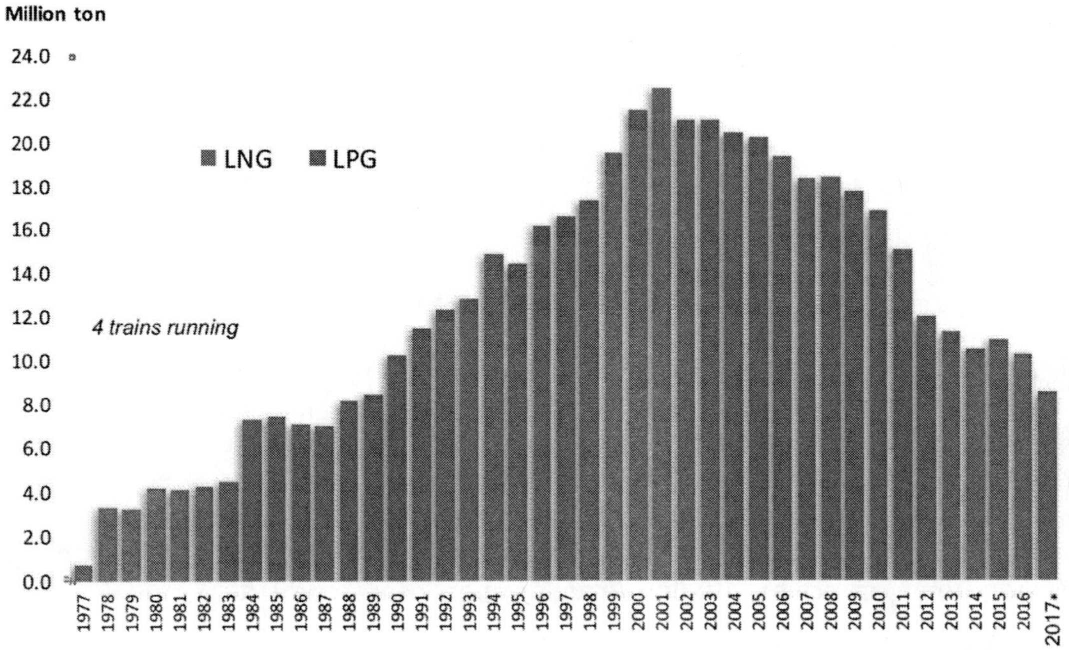

Figure 1. History of LNG & LPG Production Rate

Aiming to maintain operational sustainability, in 2017, Badak LNG had committed to receive and handle new potential feed gas resource from Jangkrik Field, East Kalimantan, Indonesia, which has methane content of >95%-mole, much higher than existing rich gas supply from Mahakam Block Field which contain only ≤87%-mol methane. This leaner feed gas emerged several operational issues that limit existing Badak LNG's plant capability to receive and handle the new potential lean gas. Therefore, process transformation had to be made by Badak LNG to overcome those issues and allow more lean gas coming to the facility.

As clearly shown in **Figure 2**, Badak LNG has started to receive leaner feed gas from Jangkrik Field in May 2017 and have successfully received about 680 MSCFD gas which equal to 90.61%-mole methane content. This paper aims to share valuable Badak LNG process transformation experiences, which includes operational adjustments and minor modifications, in attempt to overcome all identified issues that may limit Badak LNG's plant capability to handle higher methane content hence limit leaner feed gas receive from Jangkrik Field.

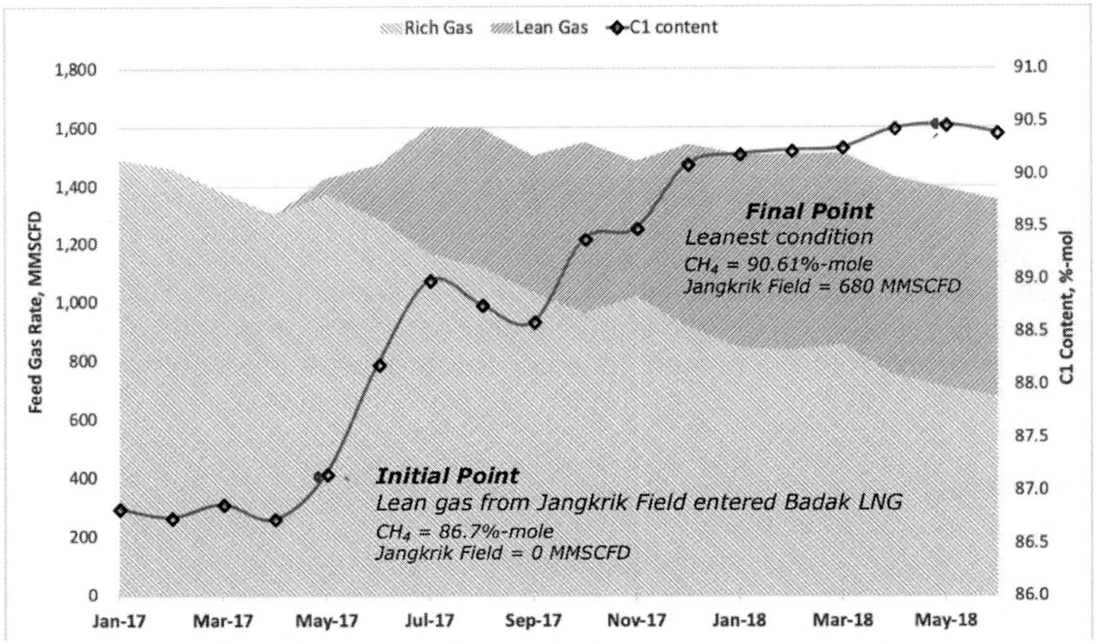

Figure 2. Leaner Feed Gas – Methane Composition Trend

Result & Discussion

In May 2017, Badak NG started receiving lean gas from Jangkrik Field with the initial point of 86.7%-mole methane content. In order to prevent undesired problem, the ramp up of lean gas supply to Badak LNG was made gradually by 10-20 MMSCFD increment per step. In each increment, Badak LNG will evaluate all key plant performances prior continue with higher lean gas supply. Several identified issues which may occur and limit the capability of Badak LNG to handle leaner feed gas discussed in this paper are as follow: (1) off-spec HHV LNG product, (2) heavy hydrocarbon carry-over, (3) low fractionation turn-down ratio, and (4) limited ethane production.

Off-spec HHV LNG Product

According to Badak LNG's sales contract, Badak LNG has responsibility to maintain density of LNG product over 453 kg/m^3 which equal to 1,107 BTU/scf of HHV LNG product. Prior lean gas entrance, Badak LNG has capability to both achieve LNG product specification and produce excess LPG product due to high propane and butane content in the feed gas. As the methane content in the feed gas increases, the amount of other heavier hydrocarbons (ethane, propane, and butane) decreased hence reduced LPG production. Once the methane content reached 88.1%-mole, no LPG production could be made since heavier hydrocarbon contained in feed gas was all used to achieve desired target for HHV LNG product. This happened when Badak LNG received approximately 157 MMSCFD lean gas from Jangkrik Field.

In order to fulfill the specification of LNG product, a project to add external LPG scheme was made. The project includes addition of LPG re-injection line from LPG tank to liquefaction plant and loading dock modification to allow receiving LPG mechanism instead, changing from previous exporting mechanism. Once completed, Badak LNG then has the capability to transfer 80 m^3/hr LPG (propane or butane) from the tanks to liquefaction plant via new external LPG re-injection line and handle lean gas up to 92%-mole methane which equal to 692 MMSCFD lean gas from Jangkrik Field without off-spec HHV LNG problem. Badak LNG has first LPG import in August 2017. The old and new re-injection mechanism are shown in **Figure 3**.

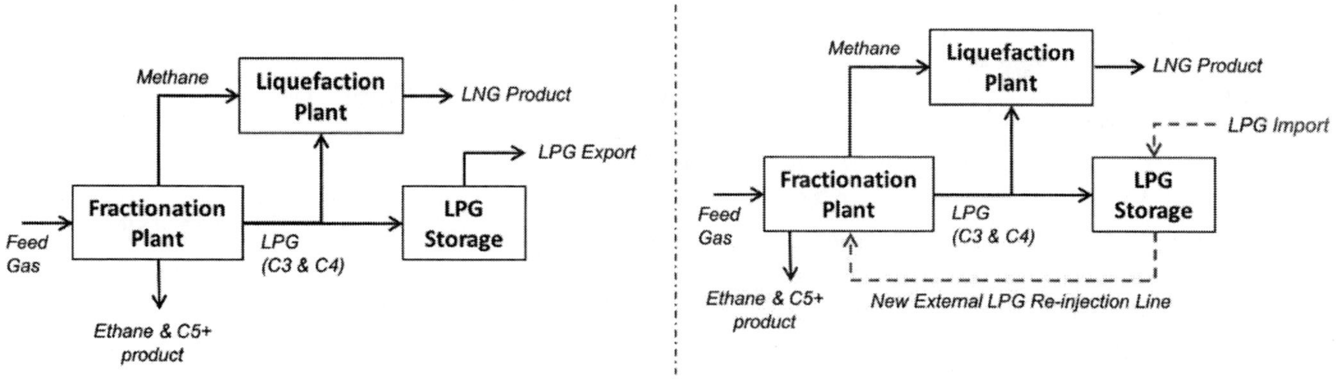

Figure 3. Old (left) and new (right) mechanism of LPG Re-injection

A careful and strict requirement must be applied to imported LPG since it may contain harmful material to the process. In November 2017, Badak LNG experienced plugging problem in liquefaction unit due to heavy hydrocarbon contamination (C_7^+) in imported LPG. This problem forced significant reduce of Jangkrik Field supply to certain rate where external LPG re-injection not required (≤ 88.1%-mole, Jangkrik Field rate = 157 MMSCFD). A comprehensive plant test is recommended to be performed prior extend usage of imported LPG.

Heavy Hydrocarbon Carry-over

Maintain heavy hydrocarbon content below its maximum limit is very important to prevent plugging problem in main heat exchanger. The content is controlled by maintaining reflux rate of scrub column and temperature of its condenser. As the methane content in the feed gas increases, the amount of other heavier hydrocarbon (especially butane) is decreasing hence reduce scrub column reflux rate. A simulation shows that the reflux rate will significantly decrease from normally higher than 26 m^3/hr to only 8.5 m^3/hr when the gas containing 89.3%-mole methane, causing significant hydrocarbon carry-over, especially i-C5 (up to 0.03 ppm, while max. allowable content is 0.01 ppm), which may cause plugging problem in liquefaction plant.

There are several modification options available to remove HHC, such as (1) front-end NGL extraction, (2) integrated scrub column, (3) adsorption, etc[1]. Taking the fastest and simplest option, Badak LNG considered external LPG re-injection to not only perform as additional LPG for maintaining HHV LNG product but also to maintain reflux rate in scrub column

[1] Khrisnamurthy, G., Liu, N. Y., "Removal of Heavy Hydrocarbons from Lean Natural Gas", presented at *AIChE Spring Meeting, San Antorio TX,* 2013.

at relatively high rate (to get over 26 m³/hr) by introducing external LPG re-injection to inlet butane recycle, in addition to re-injection scheme as shown in **Figure 4**. Once the reflux rate could be maintained at nearly 30 m³/hr, the heavy hydrocarbon carry-over could be well prevented. Except for the contamination case of imported LPG, Badak LNG has never got any issue regarding plugging problem due to hydrocarbon carry-over from scrub column up until now. Typical composition of heavy hydrocarbon in Badak LNG is shown in **Table 1**.

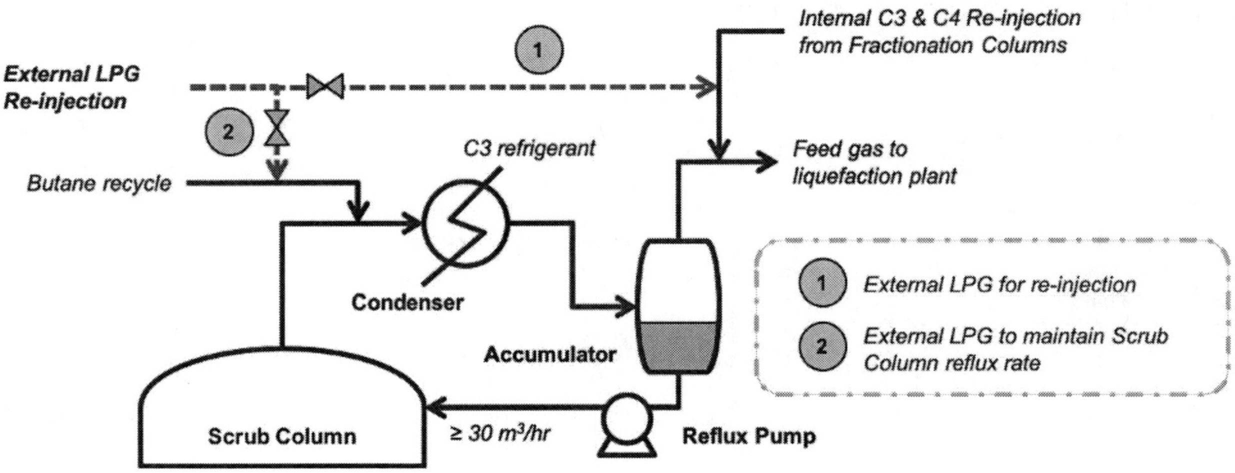

Figure 4. *External LPG to maintain scrub column reflux rate*

Table 1. Typical HHC content in Badak LNG Plant

Component	Inlet Feed Gas	Scrub Column Top Product
i-C_5	0.20%	0.02 ppm
n-C_5	0.14%	0.01 ppm
C_6^+	0.33%	-
BTEX*	400 ppm	0.20 ppm

** Benzene, Toluene, Ethylbenzene, o/m/p-xylene*

Low Fractionation Turn-down Ratio

Turn down ratio is a condition of lowest feed that could be handled by the fractionation columns. By operating at higher turn down ratio, we expecting good operational of the column in terms of separating product which meet specification and avoiding operational columns i.e. flooding, entrainment, wiping, etc. With lower methane content, available C2, C3 and C4 product will cause significant reduction of column load up to 40%. However, instead of modifying turn-down ratio of the column by closing certain amount of active area of the plate, the additional of external re-injection could significantly help the situation and make the effect of low turn turn-down ratio cause by leaner feed gas became less significant.

Limited NGL Production

In the end of 2017, Badak LNG was experiencing low NGL production, especially ethane, due to leaner feed gas. When the methane content reached 90%-mol, the inventory of C2 was lower than 50%, indicating less C2 extracted than the consumption. C2 inventory level shall be maintained at high level to fulfill C2 refrigerant make-up for either running trains or initial C2 make-up for start-up requirement. Otherwise, the operational of liquefaction in running trains may become very inefficient, the LNG product gets warmer, potential for extensive flaring due to tank overpressure and canceled train start-up.

Actually, more than 90% of the ethane contained in feed gas goes to liquefaction unit as scrub column top product instead of extracted in fractionation columns. Thus, in order to increase ethane extraction, colder scrub column operation was employed by Badak LNG. The temperature inlet of scrub column was decreased from normally warmer than -30°C to nearly -34°C by optimizing existing inlet scrub column cooler. Colder scrub column operation allows the ethane and propane to be extracted as scrub column bottom product. Of course, this effort would generally lower the feed for scrub column reflux. As previously mentioned, the presence of external LPG re-injection solved this issue. Consequently, more external LPG re-injection is required when optimizing ethane extraction. The effect of colder scrub column operation to ethane extraction were indicated by slight ethane content decrease in scrub column product as shown in **Figure 5**.

In the other hand, common method to increase ethane production such us (1) scrub column reflux by-pass or (2) reduction of condenser temperature[2] are not currently applicable in Badak LNG. The use of scrub column reflux by-pass has high potential to cause gas blow-by problem which harm the fractionation column operational stability. While reduction of condenser is design-limited which is not possible without any significant modification.

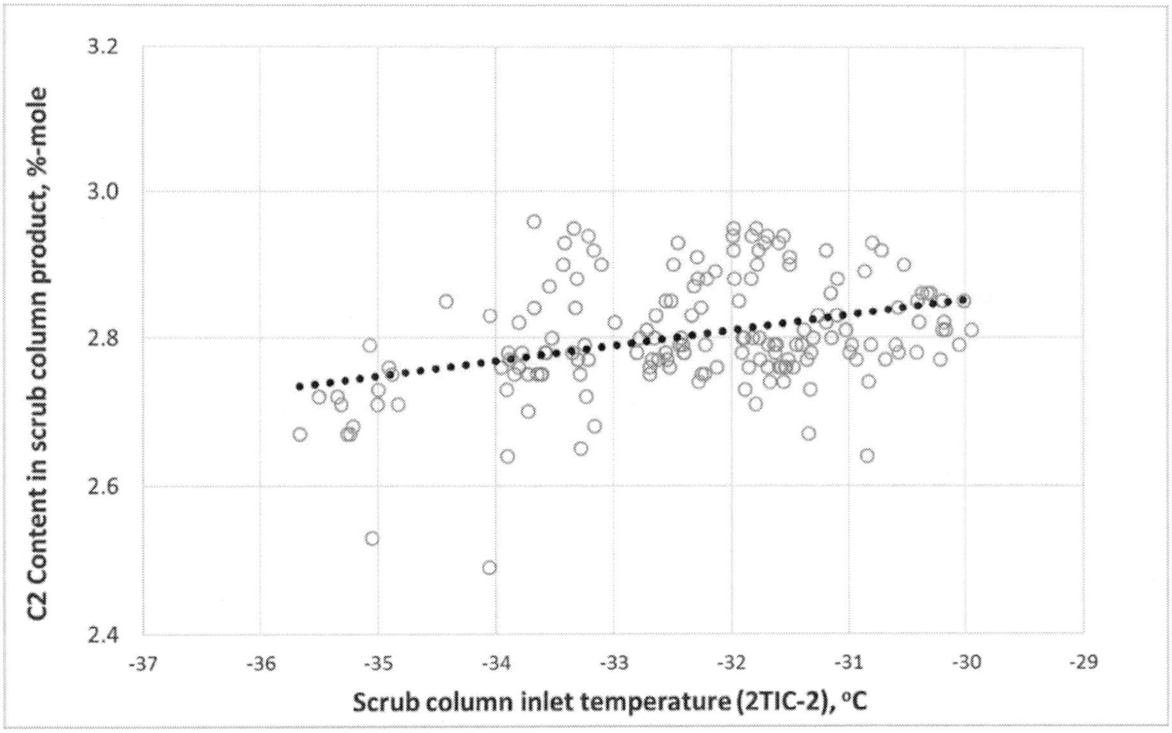

Figure 5. *Ethane content in scrub column product*

[2] Shukri, T., et al., "*Foster Wheeler outlines its latest process for handling feed-gas for the production of LNG*", LNG Journal, May 2012.

Colder scrub column has proven to increase ethane extraction in Badak LNG plant almost five times higher compared to previous condition, hence, now Badak LNG has no longer issue with limited NGL production, especially ethane, at 90.61%-mol which equal to 680 MMSCFD lean gas supply from Jangkrik Field, the highest methane content has ever received by Badak LNG so far. It was predicted that by this improvement, Badak LNG has potential capability to even cope leaner feed gas in the future without limited ethane issue. Badak LNG process transformation is summarized in **Figure 6**.

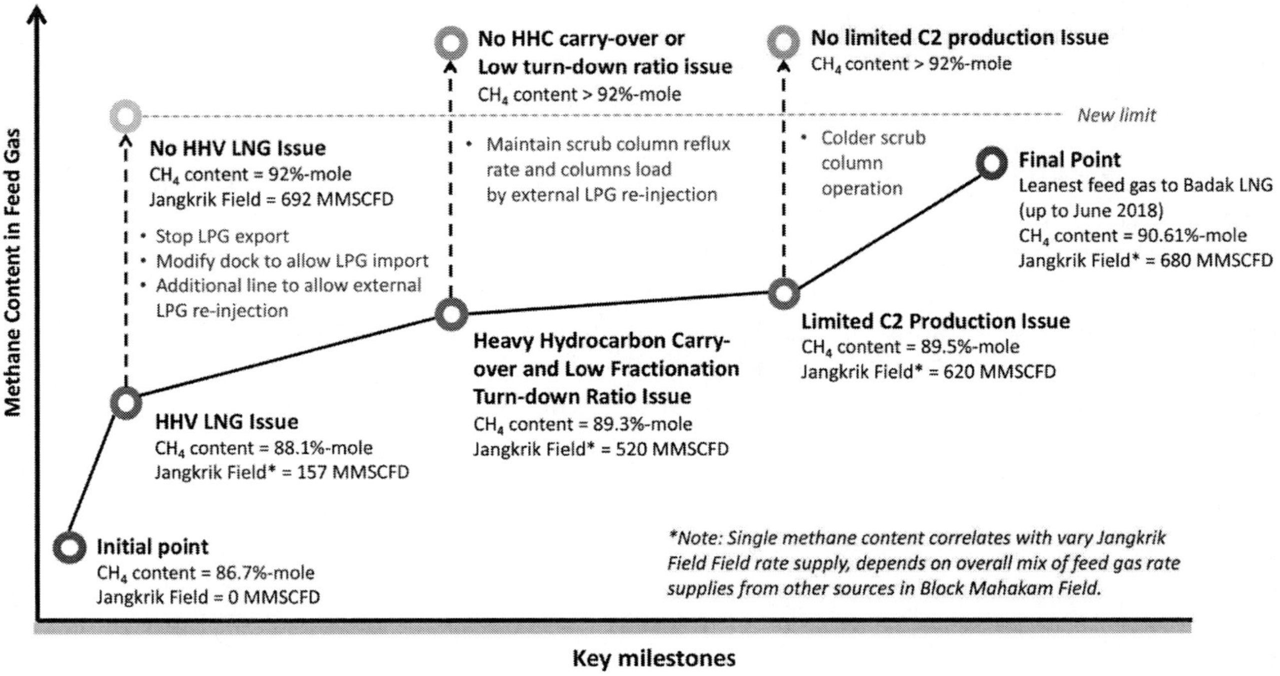

Figure 6. Badak LNG Process Transformation Summary

Conclusion

In brief, significant process transformations have been successfully made by Badak LNG to allow feed gas with higher methane content, hence maximize lean gas receive from Jangkrik Field to maintain its operational sustainability. Up to mid of 2018, Badak LNG plant had greatly handled feed gas with 90.61%-mole of methane content which equal to 680 MMSCFD of Jangkrik Field supply or 1.7 times of single running train capacity without any significant operational issue. Badak LNG even has potential capability to cope with leaner feed gas in the future.

References

1. Khrisnamurthy, G., Liu, N. Y., "Removal of Heavy Hydrocarbons from Lean Natural Gas", presented at *AIChE Spring Meeting, San Antorio TX,* 2013.
2. Shukri, T., et al., "*Foster Wheeler outlines its latest process for handling feed-gas for the production of LNG*", LNG Journal, May 2012.

NATURAL GAS LIQUIDS (NGL): FRACTIONATE OR NOT TO FRACTIONATE AT THE GAS PROCESSING PLANT

Bijal V. Gangar, Texas A&M University-Kingsville, Kingsville, TX, USA
Ali A. Pilehvari, Texas A&M University-Kingsville, Kingsville, TX, USA

Abstract

Natural Gas Liquid (NGL) is the liquid product obtained from the bottom of the demethanizer in a gas processing plant. It consists of ethane, propane, butanes and natural gasoline (heavier hydrocarbons). Here in USA, the gas processors in the planning stages of a new gas plant, must decide whether they should include the fractionation of NGL at the plant and sell the individual products or send the NGL to a central processing facility such as the one at Mont Belvieu, Texas. Overseas, most of the large gas plants fractionate the NGL at the plant site. However, in USA the gas processors must consider the economic feasibility before making such decisions, which is determined by supply and demand relationship and availability of pipelines to transport the NGL and/or NGL products. In fact, most of the old large gas plants in USA have the NGL fractionation facility but many recent gas plants have been designed without the NGL fractionation units. This paper investigates the economic feasibility or unfeasibility of having the NGL fractionation train at gas plants. The gas plants can operate in either ethane recovery or ethane rejection mode.

In case of ethane recovery, most of ethane is recovered from the bottom of the demethanizer with NGL. However, in case of ethane rejection significant amount of ethane is allowed to escape with the residue gas by warming up the demethanizer slightly. Decision whether to recover or reject ethane is made based on the prices of sales gas and ethane. If the price of sales gas is high and the price of ethane is low, then ethane rejection would be more economical.

There are many parameters that affect the economics of whether to fractionate at the plant site or not. Amongst the important parameters are price of sales gas, NGL and NGL products, cost of fractionation, transportation and storage of NGL and fractionation cost of NGL at the hub. However, in this work only sales prices, costs of fractionation and transportation are considered.

HYSYS simulation of NGL fractionation units was carried out for typical shale gas composition. Three different flowrates (150, 200 and 250 MMSCFD) were considered and the cases were analyzed in both ethane recovery and ethane rejection modes. The total module cost, operating cost and the resulting economic feasibility was analyzed for both current prices and other assumed reasonable prices of NGL, sales gas and NGL products.

Incremental economic analysis approach was utilized meaning, NGL market value at the plant site was considered as a raw material cost for the case of fractionation. For the gas composition considered and the current level of prices, the results are as follows. For both ethane recovery and ethane rejection modes at current prices, NGL fractionation at the plant site is not economical.

Introduction

Natural Gas Liquids (NGL) refers to ethane, propane, n-butane, i-butane and C_5^+ (natural gasoline). Demethanizer column separates its feed into residue gas and NGL. Residue gas, which is mainly methane, is further compressed and sold as sales gas based on its heating value. However, NGL derived from the bottom of the demethanizer column is pumped to deethanizer column where it is fractionated to ethane and C_3^+. C_3^+ is fed to the depropanizer column and split into propane and C_4^+. C_4^+ is then fed to the debutanizer column where it is separated into butanes (n-butane and i-butane) and C_5^+

(natural gasoline). The butanes can be further fractionated to i-butane and n-butane but this is commonly not done at gas processing plants.

Besides, gas processing plants are operated in either ethane recovery or ethane rejection mode based on the season, demand of ethane and price of the residue gas. A few NGL recovery process schemes have been described in the literature based on their inlet conditions, product requirements, mechanical constraints and operational requirements. Most common schemes among them are ISS (Industrial Standard Single-stage process), GSP (gas sub-cooled), CRR (cold recycle residue), RSV (recycle-split vapor), IPSI-1 (enhanced NGL recovery process) and IPSI-2 (internal refrigeration for enhanced NGL recovery process) [1]. Additional ethane recovery schemes and a new process scheme have been proposed [2] to perform a techno-economic analysis of each scheme.

The main focus of this work is to decide if one should fractionate NGL at the gas processing plant or send it to a central fractionating facility such as Mont Belvieu. However, in resolving such issues one must also consider a few factors. However, in this study only costs of fractionation and transportation have been considered. NGL can be transported via pipelines to the fractionation hub. NGL products can also be transported via pipelines from fractionators to storage and from storage facilities to end users or secondary storage [3]. Transportation fees of NGL from various basins to Mont Belvieu, ranging from 5 cpg (cents per gallon) for Permian basin to 21 cpg for Bakken field have been delineated [4]. Further, transportation and fractionation (T&F) costs of NGL has ranged from 35 cpg for old contracts to 55 cpg for newer contracts [5].

The objective of this work is

- To simulate the NGL fractionation units for different flowrates using both ethane recovery and ethane rejection modes.
- To determine the economic feasibility of NGL fractionation at the gas processing plant site.

The variables considered in achieving these objectives are flowrate of gas, current and other assumed reasonable price of residue gas, NGL and NGL products.

Process Description

The process scheme used in this work is similar to sub-cooled process (GSP) scheme, where maximum heat is recovered from the process streams and need for external refrigeration streams is eliminated. The cold stream from the bottom of the tower are withdrawn and warmed up by exchanging heat with the stream going to the cold separator. Both JT (Joule-Thomson) valve and turbo-expander have been used to provide the necessary chilling to the rectifying section of the demethanizer.

Feed gas is sweetened and dehydrated and the gas flowing from the molecular sieve unit is split into two parts. One stream is fed to the HX-2 where it is cooled by overhead stream from the demethanizer column. While, the other stream is first cooled in the HX-1 and then further cooled in HX-3. These two split streams are then mixed together and fed to the cold separator as depicted in Fig. 1.

The cold separator flashes this stream into vapor and liquid streams. The vapor stream is further split into two parts. One part is cooled by pressure reduction in turbo-expander and fed to the demethanizer. While, the other part is cooled in LNG exchanger-1, passes through JT valve where its pressure is dropped and fed to the demethanizer as top feed. The liquid stream coming out from the cold separator also passes through JT valve where its pressure and temperature are dropped before being fed to the demethanizer column.

The overhead stream of the demethanizer provides necessary cooling for the HX-2, compressed, air cooled and sent to the sales gas pipeline. The two streams from the stripping section of the demethanizer provide the cooling for HX-3, warmed up and recycled back to the column.

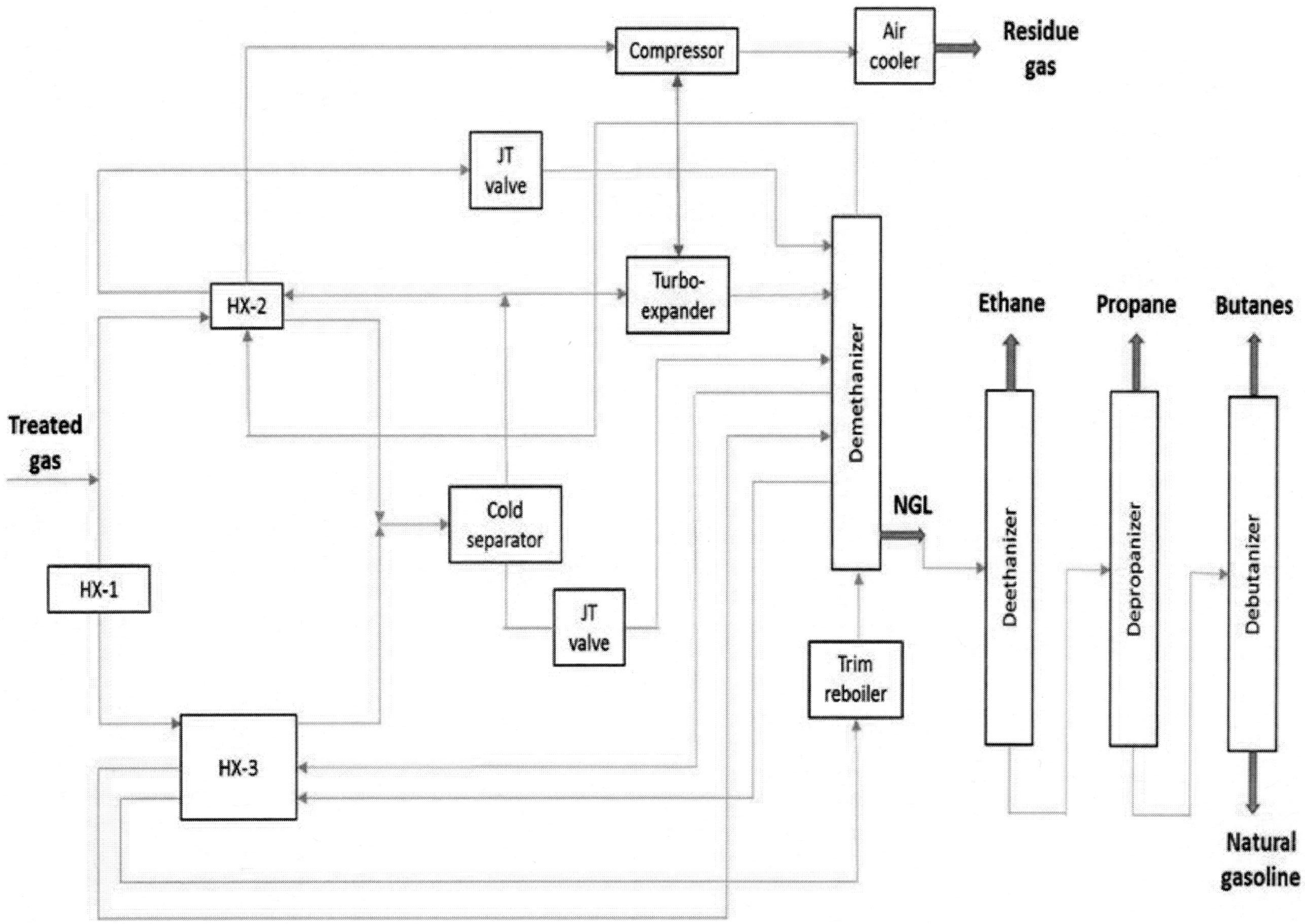

Fig. 1: Block diagram of NGL fractionation in ethane recovery mode

NGL is withdrawn as a liquid product from the bottom of the demethanizer. This way maximum ethane is recovered as demethanizer trim reboiler is turned off and column bottom is cold. Also, about 3% of methane is allowed in NGL stream [6]. This option is preferred when demand for ethane is high and the price of ethane is higher than in sales gas (ethane recovery mode). All fractionation column conditions are chosen according to the GPSA specifications [6].

The only difference in ethane rejection mode being cooling for HX-3 is provided by liquid stream from the cold separator which is then fed to the column. No side streams are withdrawn from the column and trim reboiler is turned on. This way column bottom section is hotter and NGL stream doesn't contain any methane. However, in this case significant amount of ethane is allowed to escape with residue gas as the overhead stream (within specifications).

HYSYS Simulation

HYSYS process simulations were developed for 150, 200 and 250 MMSCFD of gas in both ethane recovery and ethane rejection modes. The feed conditions and composition are rendered in Table 1. The feed compositions used are from an existing gas plant in South Texas. The composition of the gas represents a typical Eagle Ford Shale Plates.

Table 1: Feed conditions and composition for both ethane recovery and ethane rejection

Stream specifications	**Feed**		
Flowrate (MMSCFD)	150	200	250
Pressure (psia)	950	950	950
Temperature (°F)	120	120	120
Components	**Composition (mol %)**	**Components**	**Composition (mol %)**
N_2	0.0030	n-C_4	0.0057
CO_2	0.0045	i-C_5	0.0019
C_1	0.8940	n-C_5	0.0013
C_2	0.0591	n-C_6	0.0013
C_3	0.0234	n-C_7^+	0.0011
i-C_4	0.0047		

Economic Analysis

The decision whether to fractionate or not to fractionate the NGL is made based on profitability of selling NGL vs NGL products. The prices for NGL, sales gas and NGL products are based on the values at Mont Belvieu, Texas. Transportation costs have been deducted from the prices at Mont Belvieu to determine the prices at the plant site area considered, i.e. Permian and Marcellus basin. The relatively recent prices of February 2017 and prices from 2012 (when the price of ethane and other NGL products were very high) of NGL and NGL products were considered [7].

Capital Cost Estimation

Capital cost was estimated using specifications of major equipment from HYSYS simulation. The process units considered in the capital cost estimation contain only the NGL fractionation train including heat exchangers, pumps, distillation towers used in de-ethanizer, de-propanizer and de-butanizer columns. The sizing information for all these units including pressure ranges, area, number of trays, diameter and height of column were obtained from HYSYS simulations. The heat exchanger duties were also found from the HYSYS simulations.

The capital cost estimation is made using total module cost technique, which is also called as InSide Battery Limit (ISBL) plant. It includes direct and indirect costs, contingency and fee costs. Direct costs include equipment, material, utilities and operating labor while, the indirect costs consider freight, overhead, administration, distribution and selling expenses and others.

The contingency and fee costs are considered to be 15% and 3% of the bare module cost respectively [8]. The cost module is evaluated for year 2017 based on the Chemical Engineering Plant and Cost Index (CEPCI=553, [9]). The construction period of one year and project life of 15 years was assumed. The annualized interest rate of 10% was assumed with taxation rate of 30%.

The prices of sales gas, NGL and NGL products used for this study are portrayed in Table 2 [7]. The transportation costs from Marcellus and Permian basin to Mont Belvieu were considered to be \$0.15 and \$0.05 /gal respectively [4]. Operating cost for each case was calculated per the procedure outlined by Turton et al. [8].

Table 2: Current and 2012 prices of NGL, sales gas and NGL products

Component	Current prices (Feb 2017)	2012 prices
Sales gas (\$/MMBtu)	3.05	3.5
NGL (\$/MMBtu)	7.5	11
Ethane (\$/MMBtu)	3.37	11.5
Propane (\$/MMBtu)	9.6	16
Butanes (\$/MMBtu)	11.38	21
Natural gasoline (\$/MMBtu)	12.5	22

Economic analysis at Marcellus basin location and current prices

The estimate of discounted Net Present Value (NPV), Discounted Cash Flow Rate Of Return (DCFROR), and Discounted Pay-Back Period (DPBP) was obtained for NGL fractionation in ethane recovery and ethane rejection mode for 150, 200 and 250 MMscfd gas flowrate. Fig. 2 represents that discounted NPV for all three cases is highly negative and with increase in flowrate, revenue losses increase indicating that NGL fractionation in these conditions is not economical. Fig. 3 depicts economic analysis at current prices and Marcellus basin location for 150, 200 and 250 MMscfd gas flowrate in ethane rejection mode. It can be observed that NPVs for all the three flowrates are highly negative.

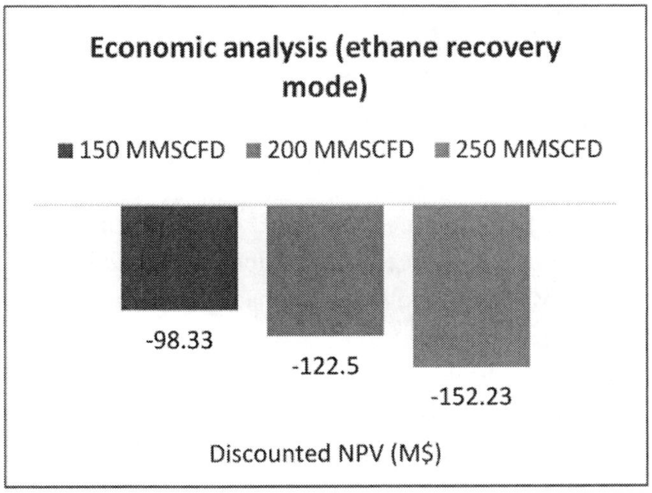

Fig. 2: Economic analysis at current prices and Marcellus basin location for 150, 200 and 250 MMSCFD gas flowrate in ethane recovery mode

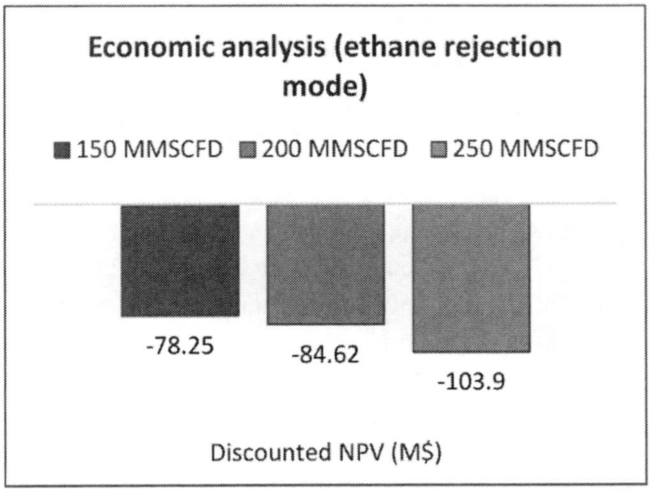

Fig. 3: Economic analysis at current prices and Marcellus basin location for 150, 200 and 250 MMSCFD gas flowrate in ethane rejection mode

However, from Fig. 2 and 3 it can be deduced that more losses are incurred in ethane recovery mode as compared to ethane rejection mode. The reason Discounted NPV is better in case of ethane rejection mode compared to ethane recovery mode is the fact that ethane as residue gas has higher value than ethane as pure. Hence, at current prices ethane rejection mode is preferred.

Economic analysis at Permian basin location and current prices

The economic analysis at current prices and three different flowrates was also performed at Permian basin location in both ethane recovery and ethane rejection modes. Even if the distance and subsequently transportation costs decreases, NPVs are still highly negative for both ethane recovery and ethane rejection modes. Which implies that, it is uneconomical to fractionate NGL on plant site in these cases.

Effect of variation in location, price of sales gas, NGL and NGL products

The economic analysis was also performed at 2012 prices. Since, there was a boom in oil and gas market at that time and it is important to check if it was economical to fractionate the NGL at the gas processing plant at such higher prices.

a. Economic analysis at Marcellus basin location and 2012 prices

Fig. 4 and 5 represents the economic analysis for 2012 vs current prices at Marcellus basin location for 200 MMscfd gas flowrate in ethane recovery and ethane rejection mode. It is evident from Fig. 4 that it was highly profitable to fractionate the NGL on plant site in ethane recovery mode with DCFROR of 154% and DPBP of 0.1 year. Nevertheless, it was uneconomical to fractionate the NGL on plant site even for 2012 prices in ethane rejection mode (Fig. 5). As ethane price was very high (Table 2) compared to residue gas and by adopting ethane rejection mode, some of the ethane is allowed to escape with the residue gas, reducing the flowrate of pure ethane.

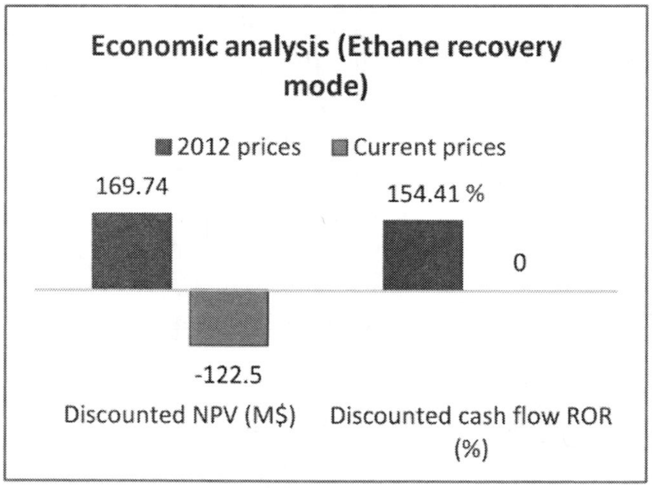

Fig. 4: Economic analysis for 2012 vs current prices at Marcellus basin location for 200 MMSCFD gas flowrate in ethane recovery mode

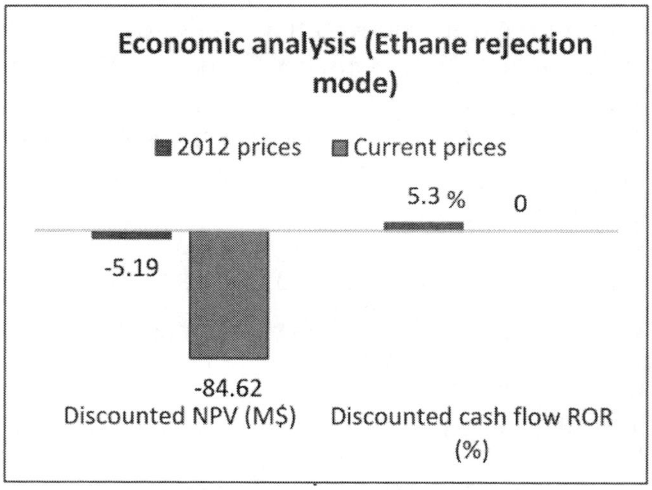

Fig. 5: Economic analysis for 2012 vs current prices at Marcellus basin location for 200 MMSCFD gas flowrate in ethane rejection mode

b. Economic analysis at Permian basin location and 2012 prices

Further, similar analysis was performed for Permian basin location for ethane recovery and ethane rejection modes and the results are similar to that for Marcellus basin location. Indicating, at 2012 prices it was economical to fractionate the NGL on plant site for ethane recovery mode but not for ethane rejection mode.

Product revenue distribution at Marcellus basin location for current and 2012 prices

The share of each product in the revenues was analyzed at both current and 2012 prices for 200 MMscfd gas flowrate and Marcellus basin location in ethane recovery and ethane rejection modes as portrayed in Fig. 6, 7, 8 and 9. This analysis helps to understand the importance of each product in revenue distribution and also identifies which product affects the revenues most significantly.

It can be depicted from Fig. 6 and 7 that since at current prices, ethane price is significantly low (Table 2) compared to other NGL products, propane plays a major role in product revenue distribution for both ethane recovery and ethane rejection modes.

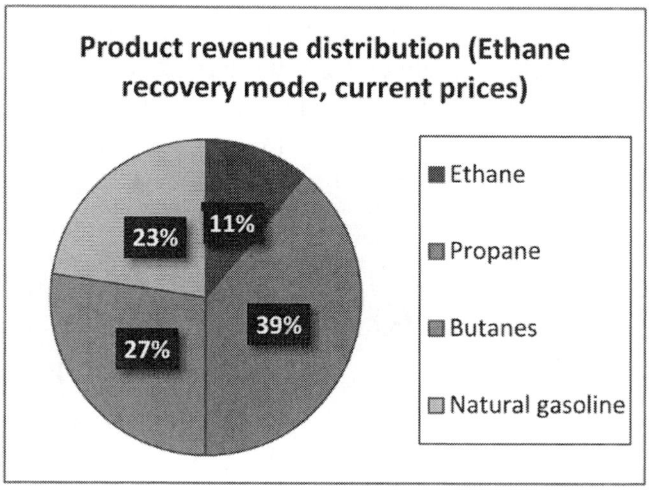

Fig. 6: Product revenue distribution for 200 MMSCFD gas flowrate at current prices and Marcellus basin location in ethane recovery mode

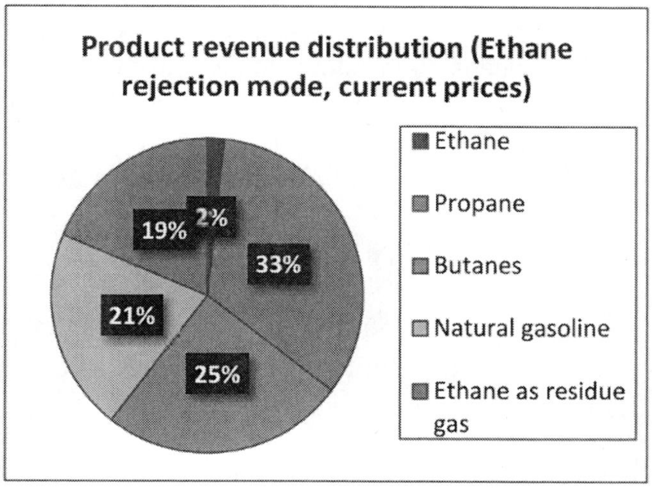

Fig. 7: Product revenue distribution for 200 MMSCFD gas flowrate at current prices and Marcellus basin location in ethane rejection mode

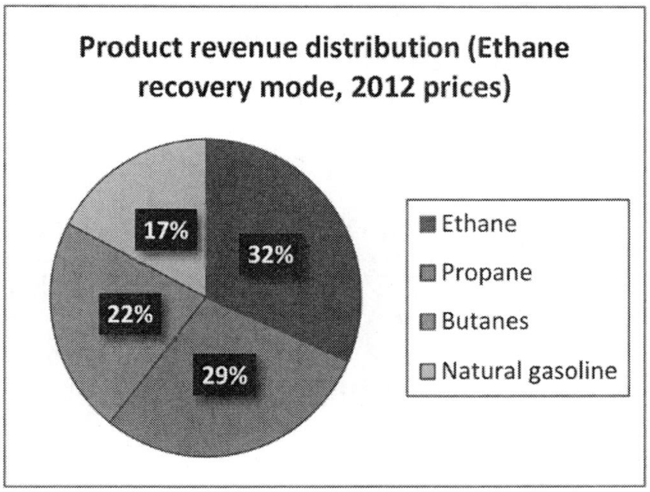

Fig. 8: Product revenue distribution for 200 MMSCFD gas flowrate at 2012 prices and Marcellus basin location in ethane recovery mode

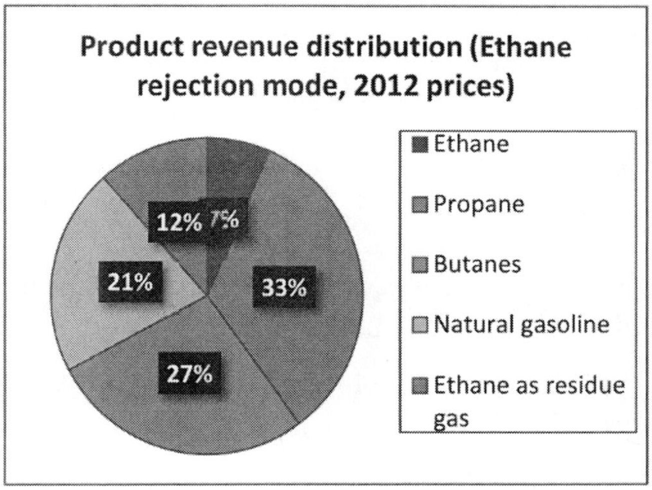

Fig. 9: Product revenue distribution for 200 MMSCFD gas flowrate at 2012 prices and Marcellus basin location in ethane rejection mode

However, in 2012 ethane prices were higher than sales gas and comparable to other products (Table 2). Thus, for 2012 prices in ethane recovery mode, ethane plays a dominant role in product revenue distribution. In ethane rejection mode at 2012 prices, again propane becomes the prevailing component as ethane flowrate decreases due to rejection of ethane in sales gas.

Conclusions

- For both ethane recovery and ethane rejection modes at current prices, NGL fractionation at the plant site is not economical. However, ethane rejection mode yields better results than ethane recovery mode.
- In contrast, NGL fractionation for 2012 prices in ethane recovery mode was highly profitable at the plant site but it was not economical in ethane rejection mode.
- Product revenue distribution is very sensitive to ethane price and in fact ethane price vs sales gas price is major factor in deciding whether to operate in ethane recovery or ethane rejection mode.
- At current prices, fractionation of NGL at the gas processing plant is not recommended.
- Further, between the prices of 2012 and 2017 there were a set of prices, during which fractionating at the plant became more economical than not fractionating it.

References

[1] Risk based optimization for representative natural gas liquid (NGL) recovery processes by considering uncertainty from the plant inlet, M. Getu, S. Mahadzir, Y. Samyudia, M.S. Khan, A. Bahadori, M. Lee, J. Nat. Gas Sci. Eng. 27 (2015) 42-54

[2] Techno-economic evaluation of a novel NGL recovery scheme with nine patented schemes for offshore applications, J.H. Park, M.S. Khan, R. Andika, M. Getu, A. Bahadori, M. Lee, J. Nat. Gas Sci. Eng. 27 (2015) 2-17

[3] Natural gas liquids (NGLs), D. Brockett, Penn state extension (2015)

[4] Consider GTL as an economic alternative for stranded natural gas and ethane, J. Cantrell, J.A. Bullin, G. Mcintyre, C. Butts, B. Cheatham, Hydrocarb. Process. 92(10) (2013) 77-80

[5] Can Mont Belvieu Handle the NGL Supply Surge? - Part II, C. Mitchell (2012)

[6] Engineering data book, Gas processors Suppliers Association, 12th edition (2004)

[7] https://www.eia.gov/energyexplained/index.cfm?page=hgls_prices

[8] Analysis, synthesis and design of chemical processes, R. Turton, R.C. Bailie, W.B. Whiting, J.A. Shaeiwitz, 3rd edition, (2009)

[9] CEPCI: http://www.chemengonline.com/current-economic-trends-cepci-january-prelim-and-december-final/

538114 Optimal Operation for LNG Ship Tank Commissioning and Loading Via Rigorous Dynamic Simulations

Abstract:

Natural gas is fastest growing energy economy. Its production and demands are increasing rapidly. Natural gas is transported as liquid over long distances. In 2017, liquefied natural gas (LNG) trade reached 293 million tonnes per annum (MTPA), as per Shell LNG Outlook 2018 report. LNG imports grew by 29 MTPA which is 30% more than what was expected. Spot cargo deliveries increased by 17% in 2017 over 2016. Natural gas and LNG production rates are expected to increase over next couple of decades.

In this work, dynamic process simulation for both LNG ship-tank commissioning and LNG loading has been performed to study the optimal procedures when new LNG ship-tanks are put to service or old LNG ship-tanks come back to service after maintenance. At the beginning, the tanks contain humid air at ambient temperature. Before LNG can be loaded, the tanks require drying, inerting, gassing up, and cooling. Drying is necessary to avoid moisture from air to condense and solidify during cooling of the tank. Air is purged using inert gases to remove oxygen from the tanks. Then methane vapors are used to remove inert gases because some inert gases like CO_2 can solidify after adding LNG and can chock valves. Cold methane is used for initial cooling of the tank. LNG liquid is sprayed for further cooling to desired temperature. Finally, LNG is loaded into the ship tanks. The LNG ship-loading is constrained by maximum allowable cooling-rate of ship-tanks, jetty boil-off gas (BOG) compressor capacity, LNG loading line capacity, and LNG level in ship-tank. Overall, LNG loading is complicated process. Dynamic simulation is helpful to study the commissioning and loading operations. The changes in the composition of material inside the tank are reported for each step in the commissioning and loading procedures. Optimization of the process to minimize time of commissioning and loading operations, emissions, and energy requirements is carried out using the Aspen Plus Dynamics simulation tool. This study would help LNG industry to operate LNG exporting terminal smoothly and in an optimal way.

531331 Greenhouse Gas Footprints of Transportation Fuels Manufactured from Natural Gas Liquids Derived from Shale Gas

Abstract:

Hydrocarbon products derived from horizontal drilling and hydraulic fracturing of shale formations (shale gas, shale derived natural gas liquids, and shale oil) have greatly expanded US production of these hydrocarbons and have made the US the worldâ€™s largest natural gas and petroleum producer. Collectively, these resources have transformed North Americaâ€™s energy landscape. Less appreciated are the transformations these shale resources are driving in chemical manufacturing. The recent growth in natural gas and natural gas liquids (primarily ethane, propane and butanes) available as feedstocks for the chemical manufacturing industry has spurred a widespread shift, in the United States, from the use of liquid, petroleum derived, hydrocarbon feedstocks in chemical manufacturing to use of natural gas liquid (NGL) feedstocks. An emerging use of NGLs may become the manufacture of transportation fuels via dehydrogenation and oligomerization, however, this use of NGLs may be limited by the environmental footprint of these transportation fuels. Access to some markets for transportation fuels is based on the lifecycle greenhouse gas footprint of the fuel (e.g., California Low Carbon Fuel Standard). This presentation will examine greenhouse gas footprints of transportation fuel production from light alkane dehydrogenation and oligomerization processes using both conventional and emerging processes. Depending on the field producing the NGLs, these footprints can be dominated by upstream greenhouse gas emissions, and so this presentation will also examine the greenhouse gas footprints of fuels manufactured from shale gas from multiple production basins.

526045 Effect of Temperature in Methane Dehydroaromatization over Mo Supported on Sulfated Zirconia Catalysts

Abstract:

Shale gas revolution is the significant increase in the production of natural gas courtesy of the advances in horizontal drilling and hydraulic fracturing. This increase has given a strong motive for research in methane conversion to higher-value endproducts, rather than flaring, power generation, or heating. Methane conversion can be divided into two routes: non-oxidative and oxidative. One of the most widely studied non-oxidative route is the dehydroaromatization (DHA) in which methane is converted into a mixture of aromatics (benzene, toluene etc.) while simultaneously producing H_2.

$$6\,CH_4 \leftrightarrow C_6H_6(g) + H_2 \uparrow \quad (\Delta G_{298K} = +433 \text{ kJ/mol}, \Delta H_{298K} = +530 \text{ kJ/mol})$$

Out of the various catalysts studied, bifunctional catalysts that possess Methane activation sites as well as Brønsted acid sites, have been the most prominent ones especially Mo supported on HZSM5/HMCM22 catalysts [1]. The DHA reaction is typically carried out around 600 – 700 °C [2] using these most widely studied catalysts.

Recently in our lab, we studied Mo supported on another support, Sulfated Zirconia (SZ) that also possesses strong Brønsted acid sites. Loading of Mo was confirmed using Raman spectroscopy and presence of acid sites was estimated using pyridine DRIFTS. This catalyst was further evaluated for methane DHA and was found to be active for this reaction. One limitation of SZ is that sulfate can start decomposing at these temperatures [3-5], which will reduce the number of available Brønsted acid sites. Hence, it is important to study the effect of temperature in the range of 600-700 °C. For this study, SZ with 5% loading of Mo was studied at three different temperatures of 600 °C, 650 °C, and 700 °C.

Here, the results show the expected increase in methane conversion with temperature, while deactivation was also rapid, apparently due to coke formation. Selectivity differed considerably. Aromatic selectivity was found to be higher for the intermediate temperature of 650 °C. This could be due to the presence of competing rates specifically that of aromatization of dimeric intermediates and rate of sulfur evolution from surface. These rates are significantly influenced by the temperatures studied here.

Figure 1: H_2 generated with time on stream at different temperatures

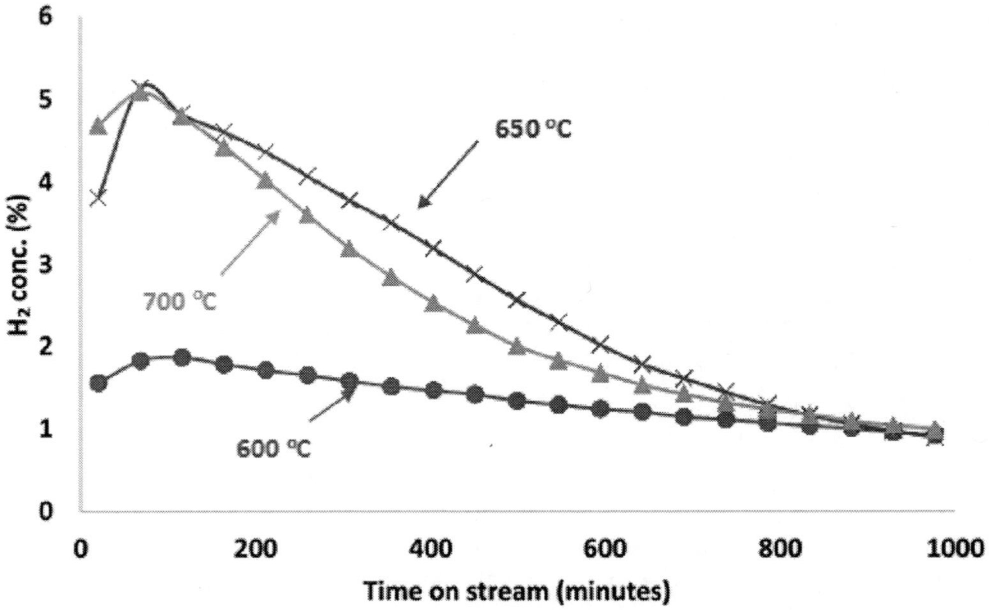

Figure 2: Time on stream selectivity towards ethylene

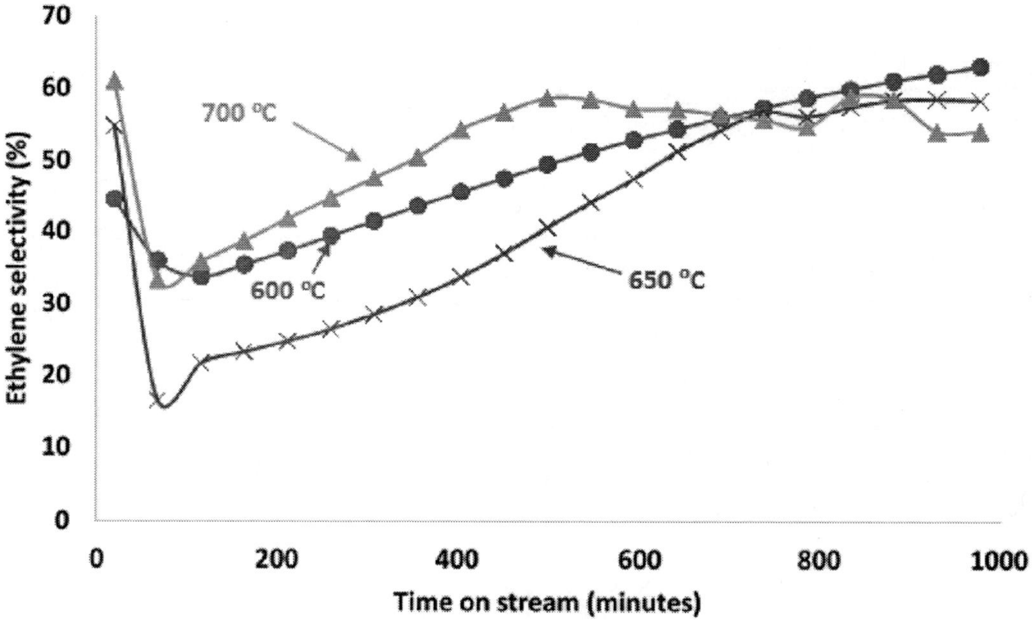

Figure 3: Time on stream selectivity towards Benzene

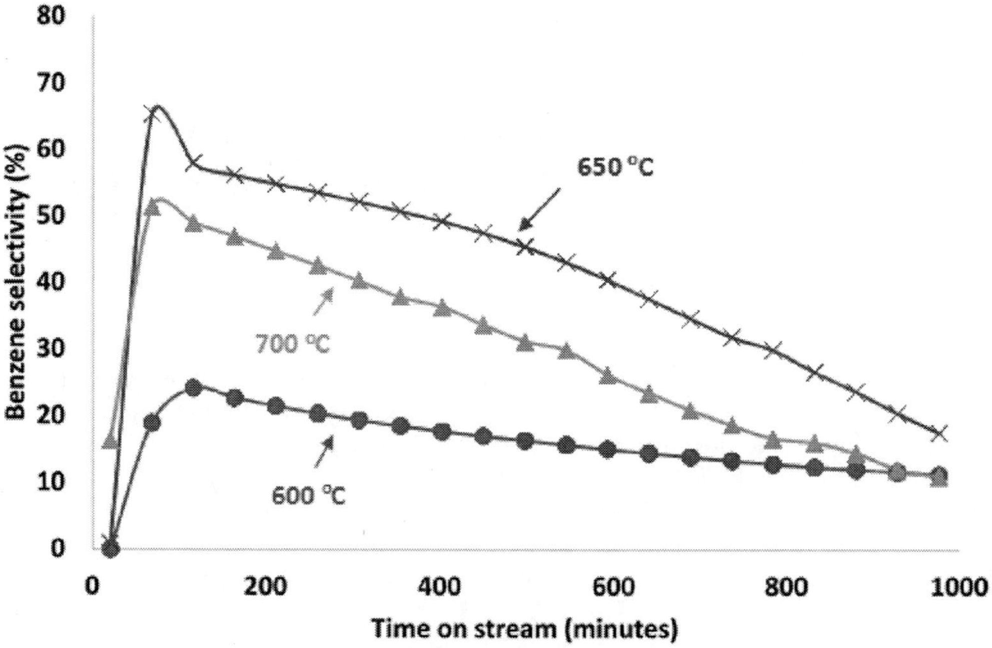

In terms of ethylene selectivity, reaction at 700 °C showed higher levels than 600 °C, or 650 °C, possibly due to higher rate of methane activation on Mo sites, which forms more dimers at the highest temperature, 700 C. At 650 °C, ethylene selectivity is lowest, possibly due to higher rate of aromatization of dimer as compared to 700 °C. Eventually all three catalysts reach similar levels of methane conversion due to coking (not shown). H_2 concentration levels also correlated well with benzene selectivity at 650 °C possibly due to faster rate of aromatization that is often accompanied by H_2 evolution.

[1] J.J. Spivey, G. Hutchings, Catalytic aromatization of methane, Chemical Society Reviews, 43 (2014) 792-803.

[2] Z.R. Ismagilov, E.V. Matus, L.T. Tsikoza, Direct conversion of methane on Mo/ZSM-5 catalysts to produce benzene and hydrogen: achievements and perspectives, Energy & Environmental Science, 1 (2008) 526-541.

[3] W.-H. Chen, H.-H. Ko, A. Sakthivel, S.-J. Huang, S.-H. Liu, A.-Y. Lo, T.-C. Tsai, S.-B. Liu, A solid-state NMR, FT-IR and TPD study on acid properties of sulfated and metal-promoted zirconia: Influence of promoter and sulfation treatment, Catalysis Today, 116 (2006) 111-120.

[4] K.M. Arata, H., Solid Superacids, Nova Science Publishers, New York, 2011.

[5] B.M. Reddy, M.K. Patil, Organic Syntheses and Transformations Catalyzed by Sulfated Zirconia, Chemical Reviews, 109 (2009) 2185-2208.

540907 Probing the Influence of Structural Changes on the Microstructural Evolution in Shale on Heating using Multi-Scale X-ray Scattering Measurements

Abstract:

Various pathways to accelerate the recovery of hydrocarbons from unconventional, nanoporous-rich subsurface environments have been proposed including the use of high temperature fluids. However, the resulting microstructural and structural changes in complex subsurface materials characterized by compositional and morphological heterogeneity are not very well understood. In this study, *in-operando* synchrotron multi-scale X-ray scattering analyses are used to determine the dynamic microstructural and structural changes in shale on heating from 30°C to 1150°C with complementary analyses of the changes in the porosity, particle size, and surface morphology. The detailed transformations of the constituents in shales including calcite, illite and quartz and the subsequent formation of denser crystalline phases such as $Al_{2.35}Si_{0.64}O_{4.82}$ (mullite), Mg_2SiO_4 (forsterite), and $Fe_{0.3}Mg_{0.7}(SiO_3)$ (enstatite) are noted from the wide and small angle X-ray scattering measurements. The changes in the roughness of the pore-solid interface on heating are determined from the combined ultra-small and small angle X-ray scattering measurements.

527210 Integration of Crude-Oil Scheduling and Refinery Planning By Lagrangean Decomposition Approach

Abstract:

Integration of crude-oil scheduling and refinery planning by Lagrangean decomposition approach

Haokun Yang, David E. Bernal, Ignacio E. Grossmann

Chemical Engineering Department, Carnegie Mellon University, Pittsburgh PA.

Problems of crude-oil scheduling and refinery planning are usually considered hard to solve to optimality at an enterprise level. The former deals with the crude oil inventory considering several types of crude oil which are delivered by different vessels and used as feedstock for the Crude Distillation Unit (CDU). The latter deals with the oil production starting at the CDU until final products involving refining and blending operations. Mixed-integer Nonlinear Programming (MINLP) models have been developed with both discrete-time and continuous-time formulation for the scheduling problem and solved with linearization based algorithms, e.g. [1] and [2]. On the other hand, the refinery planning often involves the pooling problem and usually consists of optimizing feedstocks, unit settings, as well as final product blending and shipping. The CDU modeling has used traditionally used approaches like fixed yield or swing cuts Linear Programming (LP) models. Recently, Alattas et al. [3] developed a nonlinear model of the CDU based on fractionation indices and integrated it in an MINLP model involving production planning, also considering the changeover of crudes [4]. Since both problems aim at an economic objective and are linked by the feedstock of the CDU, it becomes interesting to solve an integrated problem which may result in a better solution than the one obtained by solving each one separately. The challenge of this approach is the time scale difference between the two problems which may result in a large-scale MINLP model.

Lagrangean decomposition is a solution method which decomposes an optimization problem with complicating constraints into more easily solvable subproblems which solved separately and iteratively converge to the solution of the original problem by dualizing those complicating constraints. This method has shown a significant advantage in solving scheduling and planning integrated problems [5]. Mouret et al. [6] integrated a crude oil scheduling model with a continuous-time formulation and a simplified refinery planning model using an Artificial Neural Network (ANN) for the CDU modeling [7]. In Mouret's work, the advantages of considering an integrated approach of the crude oil scheduling and refinery planning problem were showed by achieving a time efficient framework for these problems using Lagranean decomposition.

The integrated refinery problem formulation and solution method using a Lagrangean decomposition algorithm are described in this work. We have studied the effect of having a continuous-time and a discrete-time formulation for the crude oil scheduling model and used several modeling approaches for the CDU. Two different kinds of comparisons have been made using the results of this work. Our model, which includes both the scheduling and planning simultaneously using Lagrangean decomposition with linking constraints for the CDU, is compared against a non-integrated approach to the crude oil scheduling and followed by the refinery planning. The results show an improvement of the economic objective function with respect to the sequential approach. Furthermore, we present a comparison of the solution time against solving the monolithically integrated model using state-of-the-art MINLP solvers, showing the advantage of using a Lagrangian Decomposition algorithm for this problem.

[1] H. Lee, J. M. Pinto, I. E. Grossmann, and S. Park, "Mixed-integer linear programming model for refinery short-term scheduling of crude oil unloading with inventory management," *Ind. Eng. Chem. Res.*, vol. 35, no. 5, pp. 1630–1641, 1996.

[2] S. Mouret, I. E. Grossmann, and P. Pestiaux, "A novel priority-slot based continuous-time formulation for crude-oil scheduling problems," *Ind. Eng. Chem. Res.*, vol. 48, no. 18, pp. 8515–8528, 2009.

[3] A. M. Alattas, I. E. Grossmann, and I. Palou-Rivera, "Integration of nonlinear crude distillation unit models in refinery planning optimization," *Ind. Eng. Chem. Res.*, vol. 50, no. 11, pp. 6860–6870, 2011.

[4] A. M. Alattas, I. E. Grossmann, and I. Palou-Rivera, "Refinery production planning: Multiperiod MINLP with nonlinear CDU model," *Ind. Eng. Chem. Res.*, vol. 51, no. 39, pp. 12852–12861, 2012.

[5] B. Brunaud and I. E. Grossmann, "Perspectives in Multilevel Decision-making in the Process Industry," pp. 1–34, 2017.

[6] S. Mouret, I. E. Grossmann, and P. Pestiaux, "A new Lagrangian decomposition approach applied to the integration of refinery planning and crude-oil scheduling," *Comput. Chem. Eng.*, vol. 35, no. 12, pp. 2750–2766, 2011.

[7] T. Gueddar and V. Dua, "Novel model reduction techniques for refinery-wide energy optimisation," *Appl. Energy*, vol. 89, no. 1, pp. 117–126, 2012.

536949 Optimal Production and Maintenance Scheduling in Olefin Plants

Abstract:

Process unit maintenance is defined as all actions appropriate for retaining or restoring the functionality of the unit to such an operating condition that the maintained unit can generally be reused. Maintenance is very important to ensure chemical plant reliability and profitability, as well as to help the plant safety and environmental sustainability. As a main type of maintenance, the preventive maintenance involves preplanned maintenance activities, which will be started, completed, and reused at the scheduled time. It is widely practiced in all chemical plants, especially for the olefin plant. Olefin plants employ multiple cracking furnaces in parallel to convert various hydrocarbon feed stocks to smaller hydrocarbon molecules, mostly ethylene and propylene. The continuous operational performance of cracking furnaces gradually decays because of coke formation in the reaction coils, which requires each furnace to be periodically shut down for decoking. Thus, decoking of cracking furnaces to restore them to original performance is an importance preventive maintenance for olefin plants.

In this paper, the simultaneous scheduling of olefin production and unit maintenance has been proposed. Two new maintenance initiatives: recycled ethane relay cracking and decoking grouping, are introduced to help improve the smart manufacturing in olefin plants. First, as ethane is a byproduct of cracking furnaces, it is separated and recycled as a feedstock of a furnace specially designated by the plant. Since some plants do not have equipment to store recycled ethane, another furnace must be picked up to crack recycled ethane temporary when the designated recycling furnace is in decoking. This kind of maintenance operation is called recycled ethane relay cracking. Second, when a furnace is in decoking maintenance, the throughput of this furnace will be zero and the total throughput of the whole furnace system flowing to the downstream separation process will suffer a sudden decrease. When the furnace is back on production after about two days, the throughput will become normal and the total throughput will suffer another sudden increase. These two kinds of throughput upsets due to furnace maintenance should be minimized for the stable operation of the downstream process. Thus, grouping the decoking operation means the decoking operation of different furnaces shall be lined up as many as possible to reduce the upset to downstream units. To maximize the olefin plant profit as elaborated by previous studies as well as to accomplish the two maintenance initiatives, a novel mixed-integer linear programming (MILP) model has been developed. It can help olefin plants obtain optimal schedules on simultaneous production and unit maintenance. The efficacy of the development has been demonstrated by various case studies.

536576 Ready-to-Use Operational Machine Learning in the Process Industry

Abstract:

In this talk, we present a machine learning approach that empowers the industrial practitioner to quickly solve their domain specific big data problems, and utilize the operational insight gained to improve safety, quality, and reduce downtime. This approach provides visibility and control in the hands of the practitioners who understand the operational processes and equipment (without requiring a team of data scientists). The practitioners feed the machine learning system with time series operational data, and based on the predictive insights and alerts, determine the corrective action to take. An important advantage of this approach is "continuous learning" - the practitioner can use new learnings and events to adapt, retrain, and refine the model online.

The application of this "data scientist in a box" approach in the process industry is presented through three case studies. The first case study describes the use of Falkonry for proactive health monitoring of critical equipment by a global industrial equipment manufacturer. The model was trained on operational data with identified fault conditions. The framework was integrated with their MES system issuing alerts to operations as fault conditions were detected. The second case study focused on early detection of a stream being off spec in a refinery. The model was trained on both normal as well as identified upset conditions across processing units that led to the stream being off spec. We show how the model was able to provide advance warning of the off-spec condition, which would have otherwise resulted in expensive storage and reprocessing costs for the refinery. The third case study highlights the application of Falkonry by a global semiconductor manufacturer for improving throughput through predictive maintenance of fab tools. Predictive maintenance of fab tools can reduce unexpected downtime and yield substantial benefits over the lifetime of the asset. For each of the case studies, details of the implementation, specific fault scenarios predicted, as well as organizational learnings from the implementation are discussed.

528664 Rolling Horizon Model for Gasoline Blend Planning Under Uncertainty in Demands Using an Updated Inventory Level Position Based on Chance Constraint Formulation

Abstract:

In this paper, a rolling horizon procedure is implemented to deal with gasoline blend planning problem under uncertainty in products demands. While most products from oil refineries are produced to meet contracted demand, there is an additional uncertain demands which refineries can satisfy to generate extra profit. Using a deterministic rolling horizon model leads to a suboptimal solution since such model fails to account for future production plan to meet additional uncertain demand. Inventory safety stocks approach can lead to an improved solution compared to the deterministic model, but it leaves open the question what is the right amount of safety stocks. In this paper, chance constrained programming is utilized to improve the inventory level position considering the uncertainty in future demands which is better than deterministic solution. A better prediction plan and products inventory levels can results in higher revenue by satisfying more demand and lower blending cost by using a more efficient blend recipes (lower cost per unit). Our stochastic model updates the production plan after every period based on the realized demand of the next period and new information obtained about future periods uncertain demands. The results shows that higher probability specification of the chance constraint for future planned periods will usually result in higher revenues but the blending cost per unit increase might have bigger impact on profit. Therefore different probability specification of the appropriate inventory level should be tested before making a final production plan.

543389 Novel Steady State Process Modeling Methodology for Pressure Swing Adsorption

Abstract:

Although rigorous dynamic simulation models are available for use in design and operations of pressure swing adsorption (PSA) units, such models involve sophisticated numerical solution of a complex system of partial differential equations. To explore the potential use of PSA as an intensified process alternative to conventional thermal separation processes, chemical engineers need a robust and practical PSA simulation model that properly captures the essential process fundamentals of PSA in steady state process simulation environments. We present a novel PSA steady state process modeling methodology that takes into consideration PSA unit capacity, cyclic steady state operations, mass and energy balances, and adsorption isotherms. In a way similar to "tray efficiency" in multistage thermal separation processes such as distillation, an "adsorption efficiency" is introduced to model rate-based adsorption processes. The relationship between "adsorption efficiency" and effective mass transfer coefficients are further examined with a rigorous dynamic PSA model. The modeling methodology is illustrated with two PSA examples: 1) nitrogen rejection from methane, and 2) carbon dioxide capture from natural gas. This novel PSA process modeling methodology should reprsent a new process modeling paradigm for chemical engineers to effectively examine the process intensification potential of PSA for their separation needs.

542620 The Impact of Shale Gas and Oil on the Chemical Industry

Abstract:

Lower alkanes and aromatics are very useful as both fuels and chemical raw materials. Occasionally supply disruptions constrained availability but also encouraged investigation of alternative feedstocks and chemistries. Recent new technologies for the production of unconventional gas and oil sources from tight shale formations significantly increased availability of both methane and condensate, radically altered process economics, resulted in the construction of much new intermediates capacity, and multiplied the ways traditional and alternative chemistries might be exploited. This presentation will outline historically how natural gas components became useful as chemical feedstocks, how government regulations occasionally led to perturbations in availability and price, how supply crises led to the development of alternative chemistries, catalysts, and ultimately the exploitation of shale resources. More specifically, will be discussed how excess coproducts, especially condensate, from increasing shale gas and oil production for energy are having significant unexpected impacts on the chemical process industries and opportunities for new process chemistry research.

541186 Sustainable Energy and Chemicals: Past, Present, and Future

Abstract:

The last 40 years has seen the U.S. energy situation vacillate from gasoline rationing and concerns over Peak Oil, to projected energy and feedstock prices that were noncompetitive in global markets for chemicals production, to the economic revolution of frack gas and light tight oil. The importance of low cost energy and feedstocks in driving economic opportunity will be discussed relative to this history, along with the importance of sustainability in decision making to gain stakeholder buy-in and acceptance. Strategies for managing career paths and competencies despite changing targets and high uncertainty will be addressed.

542618 Disruptions: What the Future May Hold

Abstract:

Disruptive forces are all around us. They change the way we communicate and work. They change the way we develop goods and services. They also change the types of products that are manufactured. Examples such as the way in which we produce and consume energy or the feedstocks we use to produce product have seen many changes over the course of the last few decades, and these disruptions are expected to continue. The chemical engineering discipline and the expectations for a chemical engineerâ€™s skills has also changed and adapted as a result of these disruptive forces. A discussion of the broad themes associated with these forces and the implications that they may have to chemical engineers in the future will be discussed.

542621 Geopolitical Factors Influencing the Evolution of the Chemical Industry

Abstract:

The basic chemicals and plastics industry is built on a simple principle: find a cheap feedstock and convert it into something more valuable. However, the right feedstock depends on place and time. Certain policies and actions of various governments have influenced the evolution and profitability of the chemical industry during the past 50 years, sometimes in surprising ways. For example, high oil prices and US government funding encouraged the development and commercialization of coal gasification technology during the late 1970s and early 1980s. This enabled China to develop a competitive basic chemical industry using coal as a feedstock during the last 25 years. Similarly, an abrupt increase in oil prices during the first decade of the 21st century encouraged the present shale gas revolution, causing a major shift in the economy of producing basic chemicals and polymers in the US. If history is our guide, then what governments choose to fund and regulate may be as important as market forces in determining the shape of our industry and profession in the future.

542622 Agility & Resilience: How to Maintain Career Competitiveness in the Changing Chemical Industry

Abstract:

The world around is changing. The shift to protectionism, global instability and price volatility of raw materials and feedstocks has the potential to wreak havoc on the chemical industry. With the revolution of shale gas in North America and the sudden drop in feedstock costs, the chemical sector in the US has reasserted its historic dominance. Learn how the role of a chemical engineer can contribute to industry growth and resilience, while adapting to allow for personal agility throughout the career life cycle.

533027 Virtual Manufacturing System for Refinery Process

Abstract:

With the rapid development of refinery industry in China in recent years, the better operation of these large-scale industrial plants has become an imperative and significant requirement. Because of the very large processing capacity (usually more than 10 millions of tons per year) and long time lag of the process, a tiny fluctuation in operation will cause great influence and maybe loss in economic profits. This leads to enormous risks if a trial-and-error method is applied on actual process aiming to find the best operation pattern. Thus, a virtual manufacturing system based on accurate mechanism models will provide a good platform for improving the understanding and operation of the process.

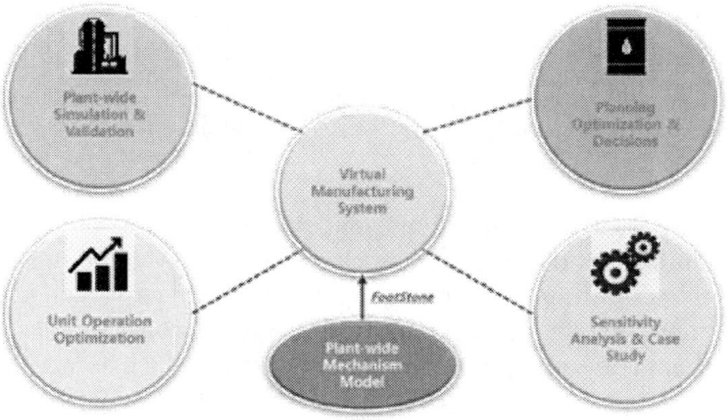

Fig.1 Architecture of the virtual manufacturing system

Taking a typical refinery factory in China as an example, the virtual manufacturing technique is developed, applied and validated. The technique structure is divided into three layers, i.e. the visualization layer, the application layer and the model layer. In the visualization layer, all the main units in the actual factory are rendered by three-dimensional modeling, which provides a virtual reality environment for real experiencing and training purpose. Through this 3D layer, the application layer is accessed for operators and engineers by clicking a specific 3D unit. The application layer consist of graphical user interface (GUI), OPC-based data communication and core application software implementing several core functions. A user-friendly GUI is developed following the DCS interface for respecting the operatorâ€™s operating habits. Then, a data coordinator is developed being responsible for data transmission between GUI and application software. This coordinator is based on industry-standard OPC protocol, which makes it flexible to be connected by a client implementing extended applications, such as the real-time optimization. It guarantees the openness of the virtual manufacturing system for application of higher level applications. In this refinery example, the application layer implements four subsystems, which are the (1) plant-wide simulation and validation, (2) sensitivity analysis and case studies, (3) unit operation optimization and (4) plant-wide production planning optimization, covering all necessary requirements from operators for improving operation level.

The architecture of the refinery virtual manufacturing system is shown in Fig.1. As the footstone for higher level applications, mechanism models of primary units in the refinery process were developed first. These models include crude distillation units, catalytic cracking, catalytic reforming, hydrocracking, hydrotreating, delayed coking and S-ZORB, etc. In the first subsystem for providing a reliable and accurate application basis, these models are running parallelly with the actual units and validating in real-time. This means the data of feedstock properties (such as the true boiling point curve, specific gravity, content of sulfur, etc.) and operating conditions (such as reaction temperature, pressure, etc.) is collected from actual process (from DCS, Real-Time Database, or Laboratory Analysis Database) and processed before inputting into the simulator and getting the simulation results. This scheme is executed every 20 minutes automatically. The simulation results and actual outputs (such as product properties, product yields, etc.) are recorded and compared. When a significant predictive bias is observed, the calibration module is triggered to calibrate key parameters of the model, which make it possible to follow and represent the change of unit characteristics. The historical comparison data can be downloaded for reproduction of the conditions and further error analysis. Based on this subsystem, the model accuracy is validated, which guarantees the reliability for the following applications, especially the optimization.

The second subsystem is for sensitivity analysis and case study. After the initialization using actual conditions at present or anytime in the past, the operator can change the operating condition or feedstock property to evaluate the effect of these changes. This will give the operator an intuitive and profound understanding of the process characteristics. Also, a reliable direction can be obtained for improving the performance of the process from current operating condition. This subsystem is especially useful for urgent adjustment of operating conditions. When some critical situations occur (such as significant change in feedstock properties), the operator may have not encountered before and had no experience in how to change the operating conditions to keep the process operating normally and profitably. Thus, this subsystem will provide a platform for the operator to evaluate different operating conditions without risks in damaging the units. A detailed report covering all input and output information of different conditions is available to the operator for further comparison and analysis. But it may be very low in efficiency when the operator wants to find an optimal operating condition with some objectives (such as economic profit of the unit) in mind, because he has to try and compare many different conditions. Thus, the next subsystem is provided for operation optimization based on mechanism models and optimization algorithms.

The third subsystem is for unit operation optimization. As the basis for optimization, an algorithm library covering stochastic algorithms, direct search methods, etc. is developed and integrated into the platform. According to the requirements of operators, all variables which can be manipulated in the unit are made available to the GUI and algorithms. One can choose several variables from the manipulating variable set to construct a user-defined optimization problem with maximization or minimization of a specific objective. There are several predefined optimization objectives (such as economic profits, total liquid yields, energy consumption, etc.) which can be chosen. A constraint set is also integrated which makes it possible to select different production modes (e.g. maximizing the yield of a specific product). With the selection of optimization algorithms, optimal operating conditions can be obtained for further evaluation and implementation. This will provide guidance for the operation of the units and is easy to make it real online implementation benefit from the openness of the system. The constraint set can also be used to take production indices from planning layer into account to find an optimal operating conditions meeting the product requirements while maximizing the economic profits.

The last subsystem is for optimal production planning which is at the highest level in the automation hierarchy and is managing global resource allocation in the factory. Traditionally, planning optimization is based on linear process model and linear programming. Because a linear process model cannot reflect the actual nonlinear characteristics of the process, this cannot guarantee the feasibility on actual process though linear programming problem is easy to solve. Thus a successive linearized model from mechanism model is included and updated in the planning optimization. The planning in this subsystem comprehensively considers the market requirements, resource supply and actual units characteristics to generate feasible and optimal planning solution. Also, to consider the uncertainties in the actual process (such as the inaccurate true boiling point curves used in the planning), a receding horizon optimization scheme is applied to update monthly plan and to divide the plan into weekly plan. This subsystem generates planning reports and guides the production in monthly scale or weekly scale. The progress of plan implementation is monitored and accessed.

The system in this case fully considers the requirements of operators and is aiming at improving the understanding and the operation levels of the refinery process. As the footstone of this system, mechanism models of primary units were developed and validated online using real-time operating data. For understanding of the characteristics of the process and some special conditions which are not frequent encountered, the sensitivity analysis and case study subsystem provides a real but undamaging operation environment. The unit operation optimization subsystem helps to find the best operation conditions of an unit considering production indices from upper level, while the planning optimization subsystem optimize the plant-wide resource allocation to maximize the global economic profits. The virtual manufacturing system is a profound practice for smart process manufacturing and is playing an important role in the refinery production.

529149 Radiative Heat Transfer Modeling Using Monte Carlo Techniques and CFD for Industrial Furnaces

Abstract:

The hottel zone method is well-established technique for analyzing radiation transfer problems in fuel-fired furnaces. However, determining the directed flux areas of complex enclosures using this method is still challenging task. In this paper, a stochastic model based on Monte Carlo (MC) method has been developed to calculate direct and total exchange areas (DEAs and TEAs) for furnaces with complex geometries. Several CFD case studies were conducted to provide valuable insight into furnace design, although they are quite computational expensive and require a fast workstation. The results of hottel zone method and CFD studies are comparable and both in a good agreement with the industrial data.

OPTIMAL CLEANING SCHEDULING AND CONTROL OF HEAT EXCHANGER NETWORKS: AN INDUSTRIAL CASE STUDY.

Federico Lozano Santamaria, Imperial College, London, UK
Sandro Macchietto, Imperial College, London, UK

Abstract

Fouling in heat exchangers of refinery applications has a high impact on operations as it reduces the thermal and hydraulic performance of the units. Fouling mitigation is required to keep the operation closer to the optimal point of low energy consumption, high throughput, and high safety standards. Two fouling mitigation alternatives are commonly used in the operation of large networks: periodic cleanings and flow rate control. The first one can remove the deposit but requires units to be taken out of service, while the second one allows a continued operation, but it just reduces the deposition rate. These two strategies have strong interactions that define the performance of the preheat train. Here we analyze the simultaneous solution of the optimal cleaning scheduling and optimal flow distribution for an industrial case study of the hot end of the preheat train. The problem is formulated as a large scale nonlinear mathematical programming problem with binary variables, which is solved efficiently using complementarity constraints. The optimal simultaneous solution of both problems has an economic benefit over the actual operation of the preheat train (based on past data) of $ 3.2 M over an operation horizon of 4 years.

Introduction

Refining operations are key in the energy sector for fuels production, but at the same time they are highly energy intensive. During crude processing large amounts of energy are required and many inefficiencies may be present in the operation. Considering the large amount of crude processed for fuels production, there are great incentives to increase the energy efficiency of refining operations which will lead to a reduction of operating cost and carbon emissions, and a safer operation. One of the main causes of energy efficiency reduction in refining operations is fouling, the deposition of unwanted material over process surfaces. Fouling decreases the heat transfer rate, increases the pressure drop in pipes, and may limit the throughput of the units in the refinery. The units require more energy for a normal operation as fouling builds up, decreasing their efficiency.

The effects of fouling are more evident in the preheat train, a large network of heat exchangers that preheats the crude before the crude distillation unit (CDU), because of the large flow rates, high temperatures, and composition of the crude. The preheat train contributes to large energy recovery in a refinery, as it may recover 65% of the energy of the product streams of the CDU [1], which for a world class facility processing 300 kbbl/day means recovering 200 MW of energy [2]. In addition, fouling is a dynamic process, as the deposit thickness increases with the operation time. To compensate for this effect, and to ensure a stable operation, the duty of the furnace has to change accordingly to the decrease in efficiency of the preheat train, such that the fuel consumption increases as the heat recovery decreases. Considering only the economic implications, a loss of 1 MW of energy recovery in the preheat train means an additional operating cost of $ 0.3 M / year [2]. Fouling of the preheat train also has environmental implications with an increase of the carbon emissions for the additional fuel consumption in the furnace and material disposal, and safety implications regarding limits of operation and dismantling process equipment during cleaning.

To compensate for the effects of fouling in heat exchanger networks different alternatives exits, but from a planning and operation perspective the focus is on two. The first one is based on the control of the flow rates through the network (via split ratios and bypasses), such that heat recovery is maximized considering the fouling resistance and heat transfer capabilities of all the exchangers. This alternative does not eliminate completely the effects of fouling, it only reduces the rate of deposition and extends the operating time of the network before it reaches operational limits. The second alternative is the use of periodical cleanings of selected exchangers, which eliminates completely the deposit present in the unit. For the duration of the cleaning the unit is out of operation, and after it is put back in the network its performance is fully restored increasing the overall energy recovery of the network. The definition of flow rate profiles through the network and cleaning schedule are not easy tasks because of the large number of interactions and variables to consider. Moreover, the two problems interact, because the optimal flow distribution in the network is no longer the best when the cleaning schedule changes.

Here a mathematical programming approach is used to formulate and solve simultaneously an optimal control problem, that defines the optimal flow rate through the preheat train, and an optimal cleaning scheduling problem, that defines the sequence and frequency of cleanings of the exchangers in a network. The integrated optimal scheduling and control problem is formulated as a large scale MINLP considering all the interaction of the network, scheduling constraints, and process dynamics. An industrially relevant case study demonstrates the ability and advantages of considering simultaneously the control and scheduling problems for fouling mitigation in heat exchanger networks.

Modelling and solution strategy considerations

The fundamental requirements for solving the heat exchanger cleaning scheduling and control problem are:

- A sufficiently accurate model to represent the performance of the network capturing the main interactions amongst the exchangers, and their individual thermal and hydraulic behavior.
- A representative model for fouling behavior. It must predict how the deposition rate changes with time, and with the main operating variables such as flow rate and temperature.
- A mathematical representation of the problem including operating constraints, and a function to measure the quality of the solution (e.g. total operating cost). This representation is expressed as an optimization problem where the degrees of freedom are the flow rates in the network and the assignment of cleanings to the exchangers at specific times.

Details of the models used are given in [6], but a short summary is presented here. Figure 1 shows a representation of a heat exchanger network, indicating all the node types. This representation is used as the basis for a general formulation independently of the number of exchangers or connections in the network. Each type of node is a subset of all the nodes of the network, and another set of arcs is used to define all the connections. In this representation the continuous degrees of freedom in a network arise from the flow split in parallel branches, while the discrete degrees of freedom are inherent to the exchanger and the time. Using this network representation mass and energy balances are formulated around each node to define the mass flow rate and temperature of each arc. For heat exchangers, the P-NTU model is used to predict the outlet temperature, and the Bell-Delaware method to define the heat transfer coefficients of the tube and shell side. For mixers and splitters adiabatic operation is assumed.

The pressure drop in the exchangers is calculated based on the friction factor and the exchanger geometry.

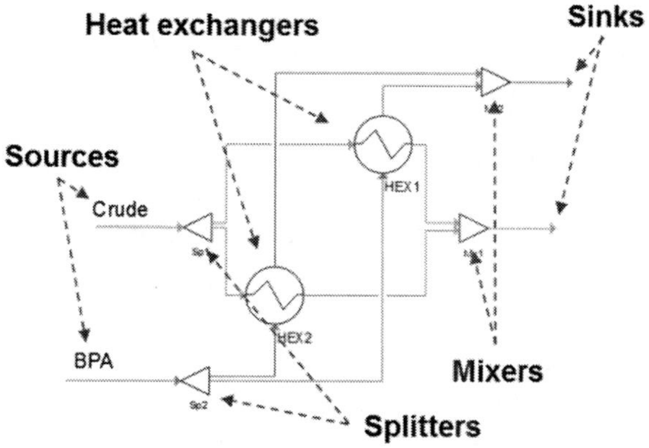

Figure 1. Types of nodes in a heat exchanger network representation

The second requirement for solving the problem is a representative model of fouling. The Ebert-Panchal model is used for this case, Eq 1, because it capture the effects of turbulence, thermal properties, and temperature on the deposition rate and also the removal of deposit due to abrasion effects or deposition suppression, determined by shear rate [3]. In addition, it is important to consider the chemical changes of the deposit as they affect its thermal conductivity. It has been shown that after a fresh layer of deposit is exposed for long times to high temperatures it changes its structure from a gel like deposit to a coke like deposit, ageing process, and this changes the physical properties of the deposit which affect the heat transfer rate [4]. The deposit ageing is modelled as a first order chemical reaction (Eq. 2), and the 'youth' of the deposit is defined as the mass fraction of gel like deposit (fresh deposit). The following set of two differential equations per heat exchanger defines the fouling model.

$$\frac{dR_{f,i,t}}{dt} = \alpha_i Re_{i,t}^{-0.66} Pr_{i,t}^{-0.33} \exp\left(-\frac{E_{f,i}}{RT_{f,i,t}}\right) - \gamma_i \tau_{w,i,t}, \quad \forall t \in Time\backslash\{0\}, i \in HEX \qquad \text{Eq. 1}$$

$$\frac{dx_{g,i,t}}{dt} = -A_{a,i} \exp\left(-\frac{E_{a,i}}{RT_{dc,i,t}}\right) x_{g,i,t}, \quad \forall t \in Time\backslash\{0\}, i \in HEX \qquad \text{Eq. 2}$$

The fouling model is described as a set of differential equations while the heat exchangers operation is modelled using algebraic equations. In addition, all the discrete decisions that define the cleaning schedule are time dependent, therefore key to the mathematical representation of the problem is how the time is discretized. This should consider both the accurate integration of the differential equations and resolution to define the scheduling decisions. Considering the trade-off between the two, a continuous time representation is used here [5], in which the time horizon is divided in periods of variable length and within each period orthogonal collocation in finite elements is used to integrate the differential equations. Figure 2 shows schematically this representation, where the length of each period is an additional decision variable for the problem. An additional advantage of this approach is that the number of binary decision variables is reduced compare to a fixed step time discretization, which is advantageous for finding a solution faster [6].

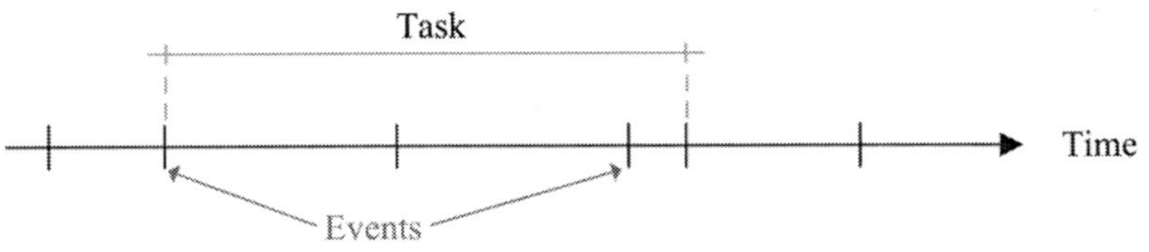

Figure 2. Continuous time discretization

The number of binary variables in the formulation, those that define which exchanger to clean and when, is the main characteristic that make the scheduling problem difficult to solve. The complexity of the problem increases with the number of binary variables because the search space becomes larger, the combinatorial nature of the problem, and the existence of multiple solutions with the same or very similar objective value. Any reduction in the number of binary variables has a significant effect on the effort for solving the problem. Although the continuous time representation reduces the number of binary variables compared to a fixed time step discretization, it is not enough to solve large scale problems. An efficient way for solving this problem of a large number of binary decisions is to introduce complementarity constraints [7]. In this case the binary variables are relaxed to be continuous variables bounded between zero and one, and slack variables are introduced to ensure that at the optimal solution the value of the original variables are at one of the bounds. Complementary constraints define a relation between two variables such that one is free, while the other is at one of its bounds. Eq. 3 shows how these constraints are defined for the binary variables (y) defining the optimal cleaning schedule for heat exchanger networks. The variables s are slack variables additionally introduced into the problem to enforce the complementarity with the y variables.

$$\begin{array}{c} 0 \leq y_{i,t,c} \perp s_{a,i,t,c} \geq 0 \\ 0 \leq \left(1 - y_{i,t,c}\right) \perp s_{b,i,t,c} \geq 0 \\ 0 \leq s_{a,i,t,c}, s_{b,i,t,c} \leq 1 \\ s_{a,i,t,c} + s_{b,i,t,c} = 1 \end{array}, \quad \forall i \in HEX, t \in Time, c \in Cleaning\ Types \qquad \text{Eq. 3}$$

The reformulation of the problem using complementarity constraints allows to find an optimal solution without exploring the whole search space (e.g. branch and bound type algorithms), but it introduces additional nonlinearities and complex relations that violates the constraint qualification of the optimization problem. This implies the problem is not well posed and cannot be solved using standard optimization solvers. To overcome this, the complementarity constraints are formulated as inequalities using a regularization approach based on an ϵ parameter, *Reg(ϵ)*, such that as $\epsilon \rightarrow 0$ the solution of the reformulation satisfies all the constraints of the original MINLP formulation [8].

Case study and problem formulation

The case study analyzed here involves the hot end of the preheat train of a refinery. Figure 3 shows the structure of the network, composed by 9 exchangers. The crude oil is split in two parallel branches which interact through the hot residue (RS) stream. All the exchangers are shell and tube type, and more details about their specifications and normal operating conditions can be found in [9]. The operation horizon is 4 years, and the fouling parameters and deposit roughness are estimated from temperature, pressure, and flow rate measurements.

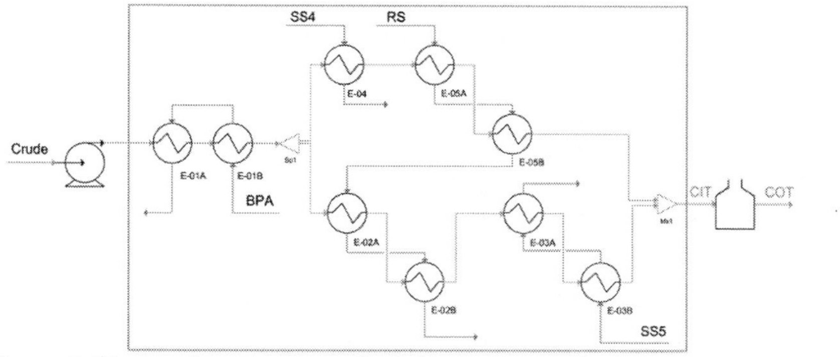

Figure 3. Heat exchanger network of the preheat train for the case study.

The fouling parameters had been determined by solving the data reconciliation problem for a more detail model based on fundamental principles [10]. However, this model is not suitable for optimization purposes due to its large scale, so a new set of fouling parameters is determined based on the prediction using the axially lumped exchanger model (P-NTU) summarised above. Relative to the more detailed fully distributed dynamic model, the prediction errors of this model on key variables are: 0.2°C for the CIT, 0.5 MW for the duty, and 0.34 bar for the pressure drop per exchanger. Those errors are within an acceptable level considering the uncertainties and the errors of the measurements. It is important to note that both models capture the important trends and interactions of the performance and behavior of heat exchanger networks.

With the fouling model and exchanger network model validated, the optimal scheduling and control problem can be formulated as follows:

$$\min_{m,y} z = P_{kg}\int_0^{t_f}\left(m_{production}^{clean} - m_{production\ ,t}\right)dt + P_{fuel}\int_0^{t_f}\frac{\left(Q_f - Q_f^{clean}\right)}{\eta}dt + P_{CO_2}m_{CO_2}\int_0^{t_f}\frac{\left(Q_f - Q_f^{clean}\right)}{\eta}dt + \sum_{i\in HEX}\sum_{c\in Cleaning\ Type}\sum_{t\in Time} P_{Cl,i,c}\, y_{t,i,c}$$

s.t. *Heat exchanger model(t,i),* $\forall t \in Time, i \in HEX$ Eq. 4

Fouling and ageing model(t,i), $\forall t \in Time, i \in HEX$

Pressure drop model (t,i), $\forall t \in Time, i \in HEX$

Time discretization constraints(t), $\forall t \in Time$

Network connectivity constraints(t,j,k), $\forall t \in Time, j \in Nodes, k \in Fluids$

Scheduling constraints (t,i), $\forall t \in Time, i \in HEX$

Operational constraints (t,i), $\forall t \in Time, i \in HEX$

Disjunctions (t,i), $\forall t \in Time, i \in HEX$

The objective function reflects to total operating cost during the time of operation and is composed by the loss of throughput cost, the energy cost, the carbon emission cost, and the cleaning cost. The constraints of the problem are: the network connectivity, the heat exchanger models, the fouling model, the time discretization, the operational constraints (e.g. limits on temperature and pressure), any scheduling constraints (e.g. partial sequence or time window for cleanings, maximum number of cleanings), and the disjunctions (e.g. to model the switch from operating state to cleaning state).

Results and discussion

The methodology and formulation presented above are used to solve two optimization problems: 1) the optimal cleaning scheduling with fixed flow split of crude oil in the parallel branches at 50%/50% (labelled "Optimal CS"), and 2) the simultaneous optimal cleaning scheduling and flow split control (labelled "Optimal Control & CS"). Both scenarios are compared against two benchmarks, the alternative of no mitigation (constant flow split and no cleanings), and the actual cleaning schedule for the network during the 4 years operation (labelled "Actual CS").

The cleaning sequence is significantly different for each of the cleaning schedules, and therefore the energy consumption is different. Figure 4 shows the furnace duty for the three schedules which is a direct measurement of the energy consumption of the process. In the actual cleaning schedule there are few cleanings and not many drastic changes in the duty. It was a rather conservative operation. On the other hand, in the optimal cleaning schedules, both with and without control, the number of cleanings and changes in furnace duty are higher. Usually more than one shell is cleaned simultaneously in the optimal solutions and this required a higher energy consumption in the furnace during the cleaning periods, because of the lack of units in the network. However, the energy recovered after the cleaning is much higher than for the actual cleaning schedule. In addition, comparing the optimal cleaning schedule with fixed flow split with that one with additional optimal flow control, in the latter the furnace duty peaks during cleaning are smaller, and the energy recover after the cleaning is higher. This shows that changing the flow rate through the network can improve the operation by taking advantage of the units with a higher capacity to transfer heat.

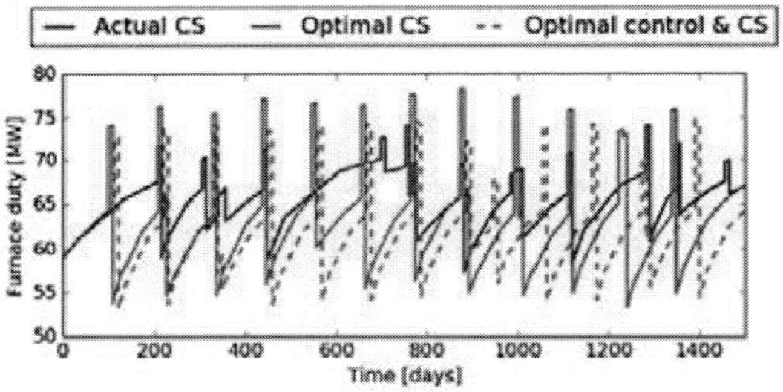

Figure 4. Furnace duty of the case study for different operating scenarios.

Figure 5 shows the three cleaning scheduling alternatives. The number of cleanings is significantly different, and the optimal cleaning schedule with fixed flow rates exhibits the largest number of cleanings, 52. Although the result of a large number of cleanings and more aggressive operation, this alternative is more profitable than the actual cleaning schedule implemented, reducing the operating cost by $ 2.0 M. Moreover, there is a strong synergy between flow control and cleaning scheduling, such that when both problems are considered simultaneously, the number of cleanings is reduced to 40 and the operating cost decreases to $ 56.4 M, which means additional savings of $ 3.2 M compared to actual schedule, and of $ 1.2 M compared to the optimal cleaning schedule with fixed flow rates. This benefit in operational cost demonstrates the synergy that exist between the flow rate distribution of the network and the cleaning scheduling, and it also shows the importance of the correct definition of the cleaning time and sequence. Fewer cleanings at the right time can be more profitable than more frequent cleanings.

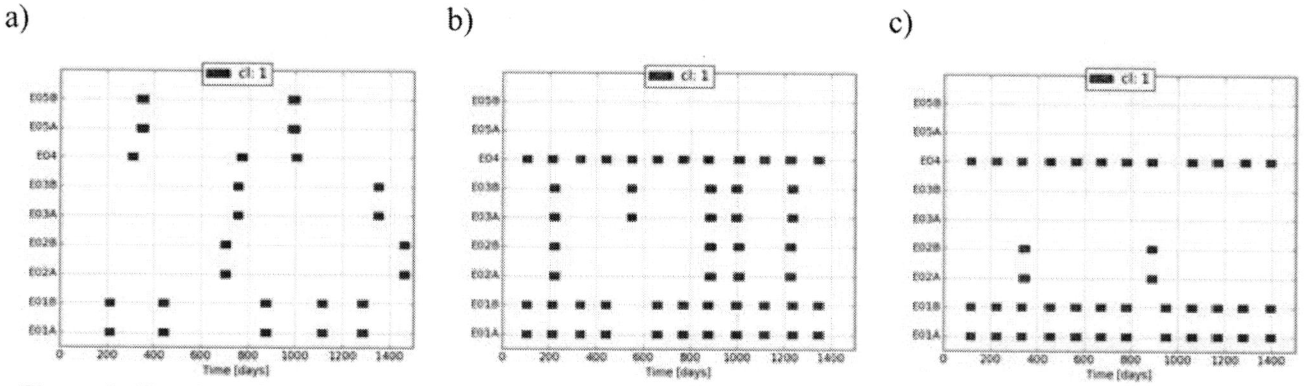

Figure 5. Cleaning schedules for a) actual operation, b) optimal scheduling with fixed flow rates, and c) optimal scheduling and optimal flow control.

For the simultaneous optimal cleaning scheduling and flow split control, there is a clear redistribution of the flow after exchangers in the branches are cleaned. Figure 5 shows the flow rates over time in the network branches including E02A and E04, respectively. The heat recovery capacity of each branch changes over time because of fouling and the cleanings, so minimal energy consumption (or minimal operating cost) is achieved by taking advantage of the branch that allows to recover more energy. The flow rate through a branch after one exchanger has been cleaned increases, while it decreases during the cleaning period, aiming to transfer the largest amount of heat to the crude oil stream while satisfying the operating conditions of the network. Note that in this case the flow through the branches is not limited by pressure, and large changes are allowed. Those changes occur both before and after the cleanings and allow a rapid distribution of the flow to increase the heat transfer.

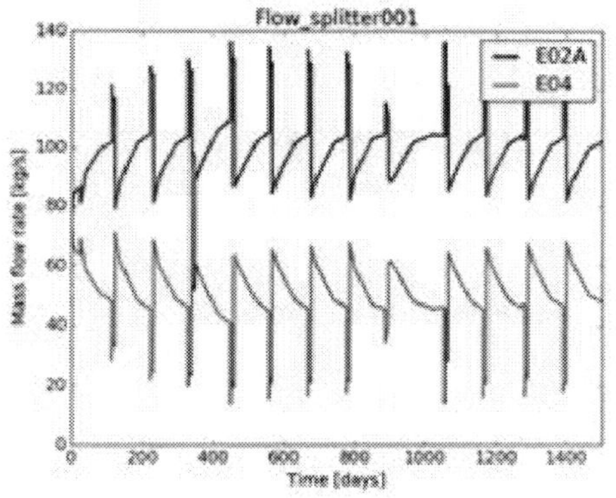

Figure 6. Flow rate distribution of crude oil between parallel branches.

Conclusions and perspectives

Optimal cleaning scheduling and control are two alternatives for fouling mitigation, and when solved simultaneously they can provide greater benefits to the operation that just implementing one or the other. For the case study of a preheat train of a crude distillation unit, the optimal cleaning schedule reduced the energy consumption and operational cost, compared to the actual fouling mitigation strategies of the refinery. Moreover, the simultaneous optimal cleaning scheduling and flow distribution improved the benefits even further, from \$ 1.2 M to \$ 3.2 M (more than doubling it) over the actual schedule. The formulation presented here is flexible, general, and in combination with the effective solution scheme discussed is able to handle large problems of industrial interest which successful results.

References

1. Yeap BL, Wilson DI, Polley GT, Pugh SJ (2004) Mitigation of Crude Oil Refinery Heat Exchanger Fouling Through Retrofits Based on Thermo-Hydraulic Fouling Models. Chem Eng Res Des 82:53–71
2. Andersson E, Tahric I (2018) Enhanced energy recovery for crude preheating. Pet. Technol. quarterly. Refin. gas Process. petrochemicals
3. Wilson DI, Ishiyama EM, Polley GT (2017) Twenty Years of Ebert and Panchal—What Next? Heat Transf Eng 38:669–680
4. Coletti F, Ishiyama EM, Paterson WR, Wilson DI, Macchietto S (2010) Impact of deposit aging and surface roughness on thermal fouling: Distributed model. AIChE J 56:3257–3273
5. Floudas CA, Lin X (2004) Continuous-time versus discrete-time approaches for scheduling of chemical processes: a review. Comput Chem Eng 28:2109–2129
6. Lozano Santamaria F, Macchietto S (2018) INTEGRATION OF OPTIMAL CLEANING SCHEDULING AND CONTROL OF HEAT EXCHANGER NETWORKS UNDERGOING FOULING: MODEL AND FORMULATION. Ind Eng Chem Res. doi: 10.1021/acs.iecr.8b01701
7. Santamaria FL, Macchietto S (2018) Integration of Optimal Cleaning Scheduling and Flow Split Control for Crude Oil Fouling Mitigation in the Operation of Refinery Heat Exchanger Networks. Comput Aided Chem Eng 44:1087–1092
8. Biegler LT (2010) Nonlinear Programming. Concepts, Algorithms and Applications to chemical process. SIAM, Society for Industrial and Applied Mathematics, Philadelphi
9. Coletti F, Macchietto S (2011) Refinery Pre-Heat Train Network Simulation Undergoing Fouling: Assessment of Energy Efficiency and Carbon Emissions. Heat Transf Eng 32:228–236
10. Coletti F, Macchietto S (2011) A Dynamic, Distributed Model of Shell-and-Tube Heat Exchangers Undergoing Crude Oil Fouling. Ind Eng Chem Res 50:4515–4533

535499 RigorousÂ Thermodynamic Analysis of a Baseload LNG Chain with Different Boil-Off Gas Minimization Strategies

Abstract:

Global energy demand is increasing and, to date, fossil fuels, such as coal, natural gas (NG), and oil, are the main energy sources. However, the combustion of this carbon based fuel is contributing to todayâ€™s environmental issues. For instance, fossil-fuel combustion in the power generation sector is responsible for nearly 22% of the worldwide greenhouse gas emissions (such as CO_2), the main contributors to global warming [1]. Having said so, for a given amount of energy, NG produces the least amount of CO_2, making it the most vital carbon based energy source. This may also explain the continuous increase in NG demand. For example, in the next decade or so, NG demand is forecasted to increase by 37% [2]. The most feasible way for the transportation of large quantities of NG over long distances is via liquefaction. This, however, requires a complex and highly interactive chain consisting of NG extraction, purification and liquefaction, storage, loading, shipment, and end user unloading, storage, and regasification. While the chain provides an environmentally friendly fuel, it releases significant amounts of CO_2 because of the energy losses embedded in the various processing steps. Our simulations show that the combustion of 3 million tonnes per annual (MTA) of LNG, a typical production rate of a baseload LNG plant, produces approximately 9.9 MTA of CO_2. The LNG chain responsible to supply this LNG will need to combust around 846 MW of Lower Heating Value fuel resulting in 0.97 MTA of CO_2 generation. Accordingly, there is a continuous need to improve and optimize existing LNG chains. Full optimization of such complex system is, however, not a straight forward technique requiring different expertise ranging from engineering/process oriented insights and skills to advanced mathematical and thermodynamics techniques. Thus, it is deemed essential to identify the chain sections that must be prioritized for optimization. This requires quantifying losses and efficiencies and one of the effective tools is exergy analysis. Identifying the areas of high exergy losses within an LNG chain emphasizes the areas where optimization is necessary.

The primary focus of the study is to identify exergy losses across a baseload LNG chain supplying more than 3 MTA of LNG. In this research, a rigorous exergy analysis was performed on an entire actual LNG chain that was simulated using ProMax® and Aspen Plus® simulation software. Exergy loss across each component of the entire chain was determined taking into consideration both physical and chemical contributions. The analysis was not limited to the main processing units, as it was extended to quantify exergy loss across the chain utility sections (i.e. power and steam generators, cooling water loop, etc.). Results revealed that the main contributor to the total exergy loss is the LNG plant utility section, accounting for 69% of the total exergy loss. Within the main LNG process, significant amounts of losses were found to occur in the liquefaction, sweetening and sulfur recovery units; corresponding for 33%, 30% and 30.4% of the plant total exergy loss, respectively, excluding the plant utility. Components responsible for the highest exergy consumption were also highlighted, with the main consumers being the compressors and their drivers, LNG storage, columns (absorbers, distillations) and heat exchangers. The contribution to the total exergy loss provided valuable insights on locations where improvements are needed to translate into environmental and energy benefits. Different optimization options were identified, rapidly evaluated, and found to be attractive for rigorous synthesis and optimization. One optimization opportunity was concerned with the LNG loss during storage, also known as boil-off gas (BOG) generation.

BOG generation takes place because of high pressure LNG flashing to storage condition, heat leaks, and other factors. The evaporation of the stored LNG means loosing portion of the compression energy provided for NG liquefaction. Nearly 12 MW of compression energy is needed to liquefy one MTA of LNG. Our simulation shows that a 1 MW loss in the liquefaction results in 3.3 MW loss in the compressors gas turbine drivers. Thus, LNG loss through BOG generation should be minimized, or effectively utilized in the chain. Different BOG minimization and utilization strategies were investigated using a combination of thermodynamic, synthesis, and simulation techniques taking into consideration loading and holding (i.e. no ship loading) operation modes. Valuable insights were obtained and utilized to recommend new practices for BOG minimization and utilization.

541565 25 by 25: Chemical Engineering in the Next 25 Years

Abstract:

In 2008, the centennial year for AIChE, we asked 25 chemical engineers for their visions of the profession from around the US and the world: Nine faculty, eight from industry, and eight post-docs and graduate students. We posed four questions related to each individual's industrial sector or area of research, asking for extrapolation, new impacts on existing sectors, new sectors, and an optional, more open-ended comment on the future of the profession as a whole.

In this 110th year of AIChE, we have revisited these questions with many of the same people. Ten years has seen many developments in our profession, in society, and in our thought-leaders' experiences. Here we report on their perspectives, comparing them to the 2008 visions.

541542 The Future of Chemical Engineering Itself

Abstract:

Chemical engineeringâ€™s future is bright because it is vital to so much of society and industry -- and its future is complicated by its rapidly expanding universe of activities. It must represent depth in core skills centered on the chemical sciences including biology and materials. At the same time, it must be transdisciplinary; it must exemplify the need and value of reaching across disciplines in borderless ways.

One way is to create and disseminate a compelling vision and strategy for chemical engineering, research, and education for the future. AIChE and the National Academies are working together to that end. In 1988, the National Academies published "Frontiers in Chemical Engineering: Research Needs and Opportunities.â€ Its committee of authors outlined a wide-ranging roadmap to turn promising research opportunities into reality, while calling for researchers to embrace new frontiers.

Since 1988, that has happened -- along with many other transformative changes like the Web, fracking, personalized medicine, and social media. A new nationwide effort, aided by international contributions, will address both the identity of chemical engineering and its future directions. Its goal is to ensure that the people in chemical engineering are primed to meet society's needs and rapid changes.

544708 Accelerating Innovation through Academic-Industrial Partnerships

Abstract:

In 2011, The Dow Chemical Company launched the University Partnership Initiative (UPI) to sponsor collaborative research at leading U.S. Universities. This 10-year commitment aims to address fundamental questions in industrial chemistry and chemical engineering with alignment to Dowâ€™s next-generation process technology and product development efforts. To date, this portfolio of projects has generated an impressive array of intellectual property, publications, and numerous examples of the collaborative developments implemented at Dow. This presentation will introduce UPI program and highlight several contributions from our collaborators and leaders in the chemical engineering profession.

540559 Maximizing Uptime, Efficiency, and Safety of Industrial Operations through Early Risk Detection

Abstract:

Achieving zero incidents while maximizing process uptime and efficiency is a key vision for operating companies. An important building block towards this vision is establishment of a proactive risk mitigation culture, supported by effective systems and a strong management. Studies have shown that most of the incidents and unexpected process failures can be avoided if the operating teams get timely information about developing risks and take preventive actions early on. While advances have been made in process monitoring and automation in the past decades, there still remains significant technological and behavioral gaps that prevent plant operations from proactively mitigating process risks.

A modern industrial plant monitors thousands of parameters, generating upwards of 50-100 million data points every day. Thanks to recent breakthroughs in machine learning and artificial intelligence approaches, today there are autonomous systems that can sift through this data and point out meaningful and timely insight. This can help operating teams ascertain process issues that are hidden in the data, long before process variables reach critical levels.

In this presentation, a new autonomous system, Dynamic Risk Analyzer^{TM} (DRA), will be introduced that points out early indicators of process issues using proprietary machine learning. The risk indicators enable operations team to respond and make required process changes, days and sometimes weeks ahead of any alarm, avoiding late stage (expensive) fixes. To realize these objectives fully, a proactive management workflow must also be in place. The presentation will discuss new initiatives and workflows developed by plant operations to implement a proactive culture. Real-life case studies from plants using DRA technology will be presented on how this culture change resulted in increased uptime, efficiency and safety of plant operations.

531309 Gaussian Processes for Hybridizing Analytical & Data-Driven Decision-Making

Abstract:

Surrogate models are widely appreciated in process systems engineering [1]. The typical setting focuses on expensive-to-evaluate, possibly uncertain functions. Examples include: modular process simulators [2], integrated gasification combined cycle processes [3], a carbon capture absorber [4], and many other applications, e.g. [5-8]. Resources are typically limited, so effective decision making requires data-efficient learning.

The data science and statistical machine learning communities typically focus on models learned solely from observed data. But chemical engineering applications may also require explicit, parametric models, e.g. modeling known process constraints, operations constraints, and cost objectives [9]. So, work has integrated semi-algebraic functions with those learned from data [10] or developed semi-physical modeling techniques [11, 12].

This presentation surveys the state-of-the-art in hybridizing analytical and data-driven decision making [13]. We consider three probabilistic modeling applications to these hybrid situations:

Design of experiments for model discrimination [14]. We bridge the gap between classical, analytical methods [15] and Monte Carlo-based approaches [16]. Classical methods may have difficulty managing non-analytical model functions and data-driven Monte Carlo approaches come at a high computational cost. We replace the original, parametric models with probabilistic, non-parametric Gaussian process surrogates learned from model evaluations. The surrogates are flexible regression tools that extend classical analytical results to non-analytical models, while providing us with model prediction confidence bounds and avoiding the computational complexity of Monte-Carlo approaches.

Multi-objective optimization [17, 18]. We make novel extensions to Bayesian multi-objective optimization in the case of one analytical objective function and one black-box, i.e. simulation-based, objective function. The resulting method has been applied to a bone neotissue application [19] and a more general test suite.

Scheduling plant operations under uncertainty. For processes with equipment degradation, we use Gaussian processes to approximate large-scale, mixed-integer optimization problems.

We close by offering a broad outlook on applying probabilistic surrogate models to chemical engineering. Statistical machine learning has recently attracted significant interest in process systems engineering [20]. Here we show that state-of-the-art research in Gaussian processes [21, 22] and probabilistic modeling more generally [23] can have a big impact on chemical engineering.

References

[1] Bhosekar A, Ierapetritou M. Advances in surrogate based modeling, feasibility analysis and and optimization: A review. *Comput Chem Eng*. 108: 250-267, **2018**.

[2] Caballero JA, Grossmann IE. An algorithm for the use of surrogate models in modular flowsheet optimization. *AIChE J*. 54(10):2633-2650, **2008**.

[3] Lang Y, Zitney SE, Biegler LT. Optimization of IGCC processes with reduced order CFD models. *Comput Chem Eng*. 35(9):1705-1717, **2011**.

[4] Cozad A, Sahinidis NV, Miller DC. Learning surrogate models for simulation-based optim-ization. *AIChE J*. 60(6):2211-2227, **2017**.

[5] Boukouvala F, Ierapetritou M. Surrogate-based optimization of expensive flowsheet modeling for continuous pharmaceutical manufacturing. *J Pharm Innov*. 8: 131, **2013**.

[6] Soepyan FB, Cremaschi S, Sarica C, and others. Estimation of percentiles using the Kriging method for uncertainty propagation. *Comput Chem Eng*. 93:143-59, **2016**.

[7] Shokry A, Ardakani MH, Escudero G, Graells M, Espuña A. Dynamic Kriging-based fault detection and diagnosis approach for nonlinear noisy dynamic processes. *Comput Chem Eng*. 106:758-76, **2017**.

[8] Tran AP, Georgakis C. On the estimation of high-dimensional surrogate models of steady-state of plant-wide processes characteristics. *Comput Chem Eng*. DOI 10.1016/j.comp-chemeng.2018.02.014, **2018**.

[9] Boukouvala F, Hasan MF, Floudas CA. Global optimization of general constrained grey-box models: new method and its application to constrained PDEs for pressure swing adsorption. *J Glob Optim*. 67(1-2): 3-42, **2017**.

[10] Boukouvala F, Floudas CA. ARGONAUT: AlgoRithms for Global Optimization of coNstrAined grey-box compUTational problems. *Optim Lett*. 11(5): 895-913, **2017**.

[11] Pearson RK, Pottmann M. Gray-box identification of block-oriented nonlinear models. *J Process Contr*. 10(4):301-315, **2000**.

[12] Cozad A, Sahinidis NV, Miller DC. A combined first-principles and data-driven approach to model building. *Comput Chem Eng*. 73: 116-27, **2015**.

[13] Boukouvala F, Misener R, Floudas CA. Global optimization advances in mixed-integer nonlinear programming, MINLP, and constrained derivative-free optimization, CDFO. *Eur J Oper Res*. 252: 701-727, **2016**.

[14] Olofsson S, Deisenroth MP, Misener R. Design of Experiments for Model Discrimination Hybridising Analytical and Data-Driven Approaches. *arXiv:1802.04170*. **2018**.

[15] Buzzi-Ferraris G, Forzatti P, Emig G, Hofmann H. Sequential experimental design for model discrimination in the case of multiple responses. *Chem Eng Sci*. 39(1): 81-85, **1984**.

[16] Vanlier J, Tiemann CA, Hilbers PAJ, van Riel NAW. Optimal experiment design for model selection in biochemical networks. *BMC Syst Biol*, 8(20), **2014**.

[17] Olofsson S, Mehrian M, Geris L, Calandra R, Deisenroth MP, Misener R. Bayesian Multi-Objective Optimisation of Neotissue Growth in a Perfusion Bioreactor Set-Up. *Computer Aided Chemical Engineering*. 40: 2155-2160, **2017**.

[18] Beykal B, Boukouvala F, Floudas CA, Pistikopoulos EN. Optimal design of energy systems using constrained grey-box multi-objective optimization. *Comput Chem Eng*. DOI 10.1016/j.compchemeng.2018.02.017, **2018**.

[19] Mehrian M, Guyot Y, Papantoniou I, Olofsson S, Sonnaert M, Misener R, Geris L. Maximizing neotissue growth

kinetics in a perfusion bioreactor: An *in silico* strategy using model reduction and Bayesian optimization. *Biotechnol Bioeng*. 115(3):617-29, **2018**.

[20] Lee JH, Shin J, Realff MJ. Machine learning: Overview of the recent progresses and implications for the process systems engineering field. *Comput Chem Eng*. DOI 10.1016/j.compchemeng.2017.10.008, **2017**.

[21] Damianou A, Lawrence N. Deep Gaussian processes. In *Artificial Intelligence and Statistics*. 207-215, **2013**.

[22] Deisenroth MP, Ng JW. Distributed Gaussian processes. In *32nd International Conference on International Conference on Machine Learning*. 37: 1481-1490, **2015**.

[23] Ghahramani Z. Bayesian non-parametrics and the probabilistic approach to modelling. *Phil Trans R Soc A*. 371(1984):20110553, **2013**.

528299 A Multi-Objective Multi-Technology (MOMT) Framework to Evaluate Various Ammonia Synthesis Processes

Abstract:

Chemical engineers are facing more challenges in process development and design today to meet the requirements in three aspects: (1) multiple process objectives have to be considered along the development such as product quality, profitability, energy efficiency, environmental impact, sustainability, safety, etc., (2) various process technologies should be assessed and compared according to the multiple objectives including the new emerging ones, (3) the short development time requires sound screening and evaluation should be made at early stage of the development.

To help designers addressing the challenges, a systematic approach named Multi-Objective Multi-Technology (MOMT) evaluation framework is proposed to facilitate the evaluation procedure. The approach streamlines the traditional development layers in terms of feed preparation, reaction, separation, heat exchanger network, utilities, etc., while introduces new evaluation factors such as logistics, life cycle analysis, sustainability, etc. Therefore, the approach would evaluate various technologies especially the sustainable new ones under the same framework.

To show the efficiency of the approach, three process technologies are evaluated including (1) traditional Haberâ €"Bosch ammonia synthesis, (2) large scale water electrolysis for ammonia synthesis, and (3) distributed wind-based water electrolysis for ammonia synthesis. An insight evaluation and sound comparison could be expected under the help of the proposed MOMT approach.

534885 Dynamic Simulations and Optimization for Chemical Plant Turnaround Flare Minimization Via Multi-Plant Material Exchange

Abstract:

Flare emissions during chemical plant turnaround operations, such as start-ups and shutdowns, emit huge amounts of NOx, volatile organic compounds (VOCs), highly reactive VOCs, and thus results in tremendous industrial material and energy losses. Therefore, flare minimization (FM) is a double win effort benefiting not only regional environmental sustainability but also industrial profitability. By far, available studies are almost exclusively focused on FM strategies for a single chemical plant. There are few systematic studies focusing on FM problems among multiple chemical plants. Actually, material exchanges (ME) among multiple chemical plants could provide an augmented degree of freedom for industrial material and energy savings and emission reductions. For example, a â€œwaste streamâ€ generated during turnaround operation from plant A could be sent to another normal-working plant B to recover potential products or reuse as the feedstock. Meanwhile, the normal-working plant B could provide assisting product streams under certain conditions (i.e. high pressure or low temperature) materials to support plant A to quickly complete its turnaround. Certainly, such ME operations involves very complicated operations and must be well designed and scheduled.

In this paper, a systemic methodology with multi-plant dynamic simulations and scheduling techniques have been employed to accomplish optimal ME operations for turnaround FMs among multiple chemical plants. To demonstrate the developed methodology, a case study for FM among three olefin plants has been performed, where Plant A is at normal-working status, plant B prepares to start-up, Plant C prepares to shut-down. Plant B could import certain flow from plant A to reduce the number of furnaces fired-up before CGC startup, further reduce flaring and saving start-up time. For Plant C, off-spec materials are vaporized and exported to different locations of plant A. Thus, some of those materials could be recovered. While applying these ME operations, different operation sequences lead diverse impacts on plant Aâ€™s production and total flaring amount. Therefore, to identify the optimal operations for not only FMs, but also ensure plant safety and operability, a series of plant-wide dynamic simulations with different schedules have been conducted. This study provides new aspect and in-depth understanding on synergies multiple plants ME operations for FMs.

535065 Study for the Optimal Operation of Natural Gas Liquid Recovery and Natural Gas Production

Abstract:

Study for the Optimal Operation of Natural Gas Liquid Recovery and

Natural Gas Production

Mozammel Mazumder, Qiang Xu*, Srinivas Palanki

Dan F. Smith Department of Chemical Engineering

Lamar University, Beaumont, Texas 77710, USA

Abstract

The increasing demand of energy sources has driven cryogenic natural gas liquid recovery plants to be more efficient but more complex as well. Most of the studies on the conventional natural gas liquid (NGL) process plants (Mehrpooya, et al., 2012; Khan, et al., 2014; Chebbi, et al., 2010). In this work, the gas subcooled process (GSP) was developed to overcome the problems encountered with the conventional NGL process. First, a rigorous simulation for a new natural gas liquid NGL liquefaction process is conducted. After that, the energy consumption roadmap is disclosed through in-depth thermodynamic analysis, where the opportunities for energy consumption minimization are identified. Based on the thermodynamic analysis, a rigorous optimization model is developed and solved for energy consumption reduction of the same NGL liquefaction process. The extraction of ethane and NGL from natural gas is generally based on external refrigeration, turboexpansion, Joule-Thompson expansion, and absorption. For the developed NGL liquefaction process, the overall optimization result shows that 936 KW of total compressor brake power is saved compare with base case, accounting for near 8.45% savings. It shows that the condenser duties are also reduced 1,046 KW which accounts for about 6.56% condenser duties consumption. The COP for base case is 1.39. It increases to 1.53 after optimized, which also indicates the liquefaction process efficiency is improved after optimization. The present work is directed to improving the economics and efficiency of the process for the recovery of NGL.

References

Chebbi, R.; Amoodi, N.S.; Jabbar, N.M.A.; Husseini, G.A.; Mazroui, K.A.; Optimum ethane recovery in conventional turboexpander process. *Chemical engineering research and design.* **2010**, *88,* 779â€"787.

Mehrpooya, M.; Vatani, A.; Moosavian, S.M.A. Optimum Pressure Distribution in Design of Cryogenic NGL Recovery Processes. *Iran. J. Chem. Chem. Eng.* **2012**, *Vol. 31, No. 3.*

Khan, M.S.; Chaniago; Y.D.; Getu, M.; Lee, M. Energy saving opportunities in integrated NGL/LNG schemes exploiting: Thermal-coupling common-utilities and process knowledge. *Chemical Engineering and Processing.* **2014**, *82,* 54â€"64.

515790 Advancing Export Terminal Technology: An Optimized Process for the Refrigeration of Cryogenic Hydrocarbons

Abstract:

As a result of the â€œshale revolutionâ€ in the United States, the availability of competitively priced chemical pre-cursor feedstocks, primarily ethane and propane, has resulted in increased grassroots construction and/or expansion of various petrochemical complexes along the gulf coast. Thanks to this build-out, domestic production of petrochemical commodities, like ethylene and propylene, is expected to greatly outpace domestic consumption. Excess US domestic production and the resulting lower prices will lead to escalating international demand for these commodities overseas; domestic petrochemical producers are looking at large consumers in China, India, and Europe to take up the slack and import this excess production. In order to export these feedstocks, refrigeration units are required to chill the hydrocarbons to cryogenic temperatures. Installation of the most efficient refrigeration process at the export terminal is vital for the terminal owner to maximize their investment potential.

This paper presents the technical design for an ethylene export terminal utilizing two (2) different refrigeration processes: the traditional open cycle (OC) and a more advanced single mixed refrigerant (SMR) scheme. The two processes were modeled in-house at Chiyoda using the generic concepts of refrigeration cycles and UNISIM Design suite 450 simulation software. Each process is optimized through adjusting the available process variables to achieve an overall efficient refrigeration process for an ethylene production rate of 100 metric tonnes per hour (mt/hr). The technical design is concluded by defining major equipment sizes, layout, and a relative cost for the units. While the technical paper will focus on ethylene, similar results can be obtained for other hydrocarbons including ethane and propylene.

539461 A Novel, Environmentally Benign Supercritical CO_2-Ethanol System to Produce High-Yield Carbon Fiber Pre-Precursor from Power River Basin Coal

Abstract:

In this research, a novel, environmentally benign supercritical CO_2-ethanol(scCO_2-ethanol) system to produce high-yield carbon fiber pre-precursor from Power River Basin (PRB) coal is studied. The binary scCO2-ethanol system shows excellent performance for producing valuable liquid tar in high yield which can be used as carbon fiber pre-precursor. The results indicated that the highest yield of liquid tar reached up to 70% from three times extractions at 350°C with initial pressure 500 psi. Different characterizations methods are employed to discover the physical and chemical properties of liquid tar as carbon pre-precursor. In addition, the light oil fraction can be used as feedstock for producing valuable chemicals.

534151 NiMo-Ceria-Zirconia Anode for Direct Gasoline-Fed Solid Oxide Fuel Cells

Abstract:

NiMo-ceria-zirconia anode for direct gasoline-fed solid oxide fuel cells

Xiaoxue Hou[a], Kai Zhao[a], Qusay Bkour[a], M. Grant Norton[b*], Su Ha[a*]

[a]Voiland School of Chemical Engineering and Bioengineering, Washington State University, Pullman, WA, 99164, USA

[b]School of Mechanical and Materials Engineering, Washington State University, Pullman, WA, 99164, USA

* *E-mail: suha@wsu.edu*

To operate the Ni-based Solid Oxide Fuel Cells (SOFCs) by feeding gasoline, the complex hydrocarbon needs to be reformed into a simple synthesis gas (a mixture of hydrogen and carbon monoxide) for the anode electrochemical oxidation reaction. In our previous research, a NiMo-ceria-zirconia catalyst was found to show high reforming activity with a coking resistance towards partial oxidation of model gasoline. When this catalyst was applied as an internal reforming layer over the conventional Ni-YSZ anode of SOFCs, the single cell displayed significantly improved initial power density and performance stability under the direct isooctane/air mixture feed condition. In addition to the high reforming activity, the NiMo-ceria-zirconia can conduct both electrons and ions. Hence, the NiMo-ceria-zirconia is expected to provide a dual-functionality for single anode layer SOFCs: firstly to internally convert complex hydrocarbons into synthesis gas with a high resistance to coking and secondly to electrochemically oxidize the synthesis gas mixture for an electric power generation.

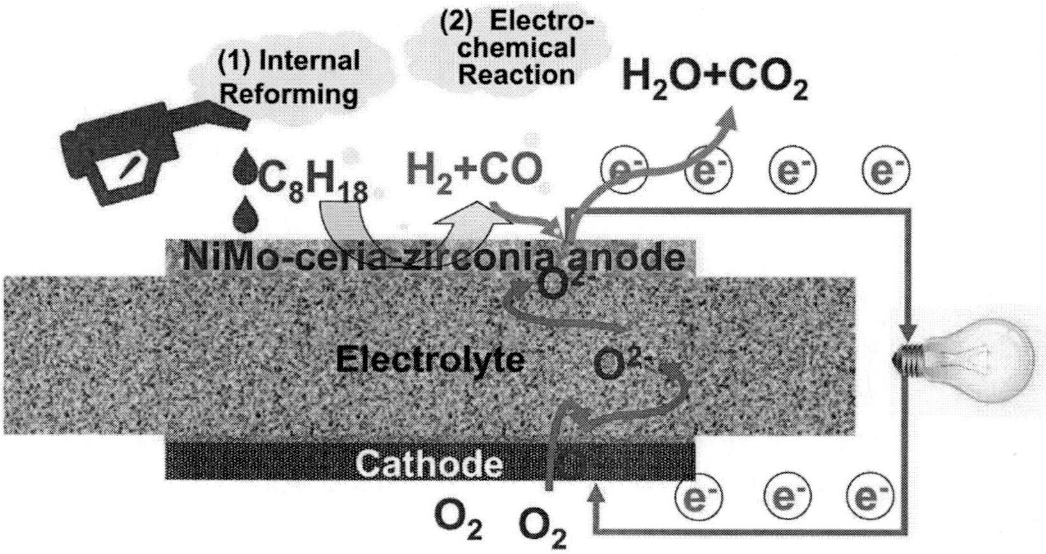

Fig. 1 Schematic diagram of SOFC single cell with the single layer anode.

In this work, we fabricated an electrolyte-supported SOFC (shown in Fig. 1) using a NiMo-ceria-zirconia single anode layer and investigated its performance under the direct isooctane/air mixture feed condition. We found that the addition of Mo to the Ni-ceria-zirconia significantly suppresses carbon deposition in the anode and improves the lifetime of the single cell. Additionally, the presence of Mo in the Ni-ceria-zirconia anode appears to facilitate a higher degree of sintering, which increases the electronic conductivity and maximum power density of the single cell. Consequently, at 800 degree C the single cell using 5 wt.% Mo in the Ni-ceria-zirconia anode displayed a higher maximum power density of 212 mW cm^{-2} at 0.48V than the single cell using the Ni-YSZ anode without the Mo promoter (49 mW cm^{-2} at 0.48 V) under the isooctane/air operation mode. Furthermore, the cell presented stable electrochemical performance with a high coking resistance over 30 h operation using the model gasoline. These promising results suggest the future application potential of the NiMo-ceria-zirconia as single layer anode for SOFCs running on complex hydrocarbon fuels.

534578 **NiMo-Ceria-Zirconia Catalyst for Inert-Substrate-Supported Tubular Solid Oxide Fuel Cells Running on Model Gasoline**

Abstract:

NiMo-ceria-zirconia catalyst for inert-substrate-supported tubular solid oxide fuel cells running on model gasoline

Kai Zhao[a], Qusay Bkour[a], M. Grant Norton[b]*, Su Ha[a]*

[a]Voiland School of Chemical Engineering and Bioengineering, Washington State University, Pullman, WA, 99164, USA

[b]School of Mechanical and Materials Engineering, Washington State University, Pullman, WA, 99164, USA

* *E-mail: suha@wsu.edu*

Solid oxide fuel cells (SOFCs) are electrochemical energy devices offering a high efficiency route for converting chemical energy in fuels into electricity. Typically, SOFCs work in a temperature range of 600-1000 degree C. The high operation temperature enables the electrochemical oxidation of not only small molecules (such as hydrogen, carbon monoxide and methane), but also complex hydrocarbon fuels, such as liquid hydrocarbon fuels and synthetic liquid biofuels.

Nickel-based cermets, such as Ni-yttria-stabilized zirconia (Ni-YSZ), and Ni-samarium-doped ceria (Ni-SDC), are widely used as anode supporters for SOFC single cells running on hydrogen because of the high catalytic activity of nickel, and a good electronic-ionic mixed conductivity of the cermet. However, when using complex hydrocarbons as fuels, a severe carbon deposition on the Ni deactivates the electrochemical properties of the anodes for the fuel oxidation, and the chemical reaction between Ni and carbon deposits leads to the structure instability of the anodes and cracking of the single cell. To address these problems, it is necessary to apply active catalysts in SOFC single cells to effectively reform the complex hydrocarbon fuels into simple synthesis gas (a mixture of hydrogen and carbon monoxide), and suppress the carbon deposition.

In our previous work, we have successfully synthesized a novel NiMo-ceria-zirconia catalyst that shows a high reforming activity toward liquid hydrocarbons (e.g., isooctane) and an excellent coking tolerance. In this work, we applied the NiMo-ceria-zirconia as an internal reforming catalyst for a tubular SOFC single cell running on the isooctane/air mixture. Fig. 1 shows the configuration of the tubular single cell. It has a porous YSZ supporter, Ni-SDC anode, YSZ/SDC bilayer electrolyte and $La_{0.6}Sr_{0.4}Co_{0.2}Fe_{0.8}O_{3-d}$ (LSCF) cathode. The NiMo-ceria-zirconia catalyst was packed into the tubular channel of the single cell as the internal reforming unit. With application of the catalyst in the tubular cell, the mixture of isooctane and air can be efficiently reformed into the simple synthesis gas (a mixture of hydrogen and carbon monoxide) at 750 degree C. Compared to electrochemical performance of the single cell without the packed catalyst, applying the packed catalyst increased the maximum power density of the single cell from 240 to 306 mW cm^{-2} using the model gasoline. Furthermore, the tubular cell with the packed catalysts exhibited significantly improved stability using the model gasoline. The results suggest the feasibility of applying the NiMo-ceria-zirconia catalyst into the tubular single cell design and operating it under the direct feed condition of liquid logistic fuels.

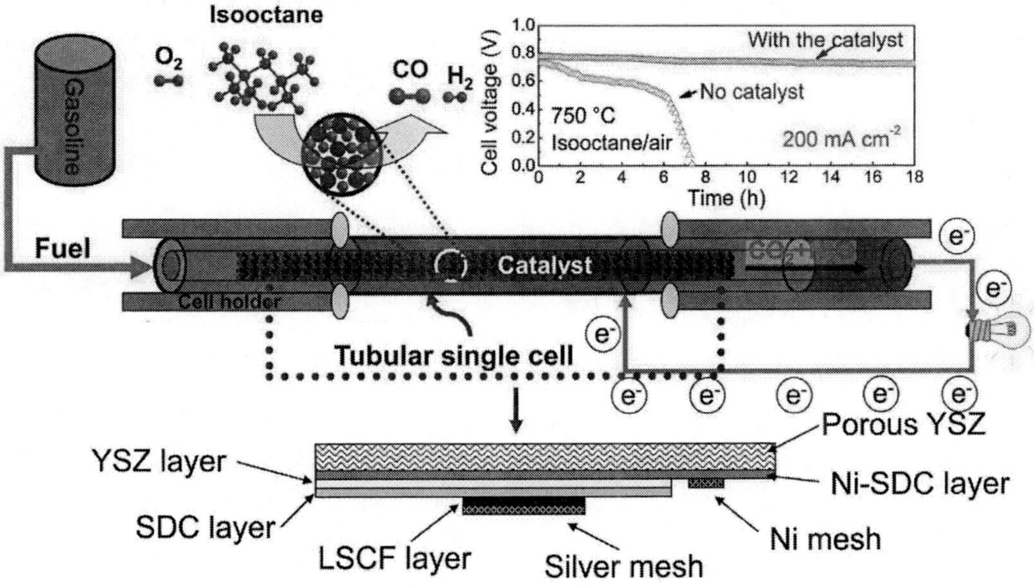

Fig. 1 Schematic diagram of the tubular single cell with NiMo-ceria-zirconia catalyst.

537198 Homogeneous Reaction Kinetics of Carbohydrates with Viologen Catalysts in Biofuel Cell Applications

Abstract:

Renewable resources, which continue to grow in utilization, still have many sustainability challenges related to economics, efficiency, supplies, and storage. Thus, continued efforts are needed to address these and other challenges to optimize the use of renewable energy. One area of exploration is the development of fuel cells, which includes biofuel cells that utilize biomass as the feedstock. This work focuses on utilizing carbohydrates, a biomass constituent, as the feedstock.

Catalysts are used in biofuel cells to enhance the energy extraction efficiency by facilitating the transfer of electrons from the carbohydrate to the anode of the biofuel cell. Based on early work in our research group, viologens have been shown to be effective catalysts in facilitating the electron transfer. However, process optimization of the catalyst is needed to provide the greatest electron transfer rates and associated efficiency. Thus, it is important to characterize how the process environment affects the rate of electron transfer.

Characterization of the homogeneous rate of electron transfer from the carbohydrate to the viologen will be presented. The heterogeneous rate of electron transfer from the viologen to the electrode will only be briefly discussed. Initial reaction rates of electron transfer from the carbohydrate to the viologen were studied in an anaerobic system as a function of pH, temperature, incubation time, type and concentration of carbohydrate (ranging from C3 to C6 carbohydrates), viologen concentration, and carbohydrate incubation time. A mechanistic model was developed to characterize the reaction rate. Model and experimental measurements showed excellent agreement. Studies were also performed to measure the rate at which different viologens decompose since decomposition can reduce process efficiencies. Decomposition rates were experimentally determined and fit to a mechanistic model. The electron transfer rates and decomposition rates were combined to show process conditions that would minimize the impact of decomposition on the biofuel cell process. With knowledge of these mechanistic models, estimates on the potential energy production of biofuel cells utilizing carbohydrates and viologens will be presented.

526602 Nature-Inspired Flow-Fields and Water Management for PEM Fuel Cells

Abstract:

Nature-Inspired Flow-Fields and Water Management for PEM Fuel Cells

Jason In Sung Cho[1,2], Toby Neville[1,2], Panagiotis Trogadas[1], Billy Wu[3], Dan Brett[1,2] and Marc-Olivier Coppens[1]

[1] Department of Chemical Engineering, University College London, London WC1E 7JE, United Kingdom, m.coppens@ucl.ac.uk

[2] Electrochemical Innovation Lab, Department of Chemical Engineering, University College London, London WC1E 7JE, United Kingdom

[3] Department of Mechanical Engineering, Imperial College London, UK

Flow-field design is crucial to polymer electrolyte membrane fuel cell (PEMFC) performance, since non-uniform transport of species to and from the membrane electrode assembly (MEA) results in significant power losses[1]. The long channels of conventional serpentine flow-fields cause large pressure drops between inlets and outlets, thus large parasitic energy losses and low fuel cell performance[2].

Here, a lung-inspired approach is used to design flow fields guided by the structure of a lung. The fractal geometry of the human lung has been shown to ensure uniform distribution of air from a single outlet (trachea) to multiple outlets (alveoli). Furthermore, the human lung transitions between two flow regimes: 14-16 upper generations of branches dominated by convection, and 7-9 lower generations of space-filling acini dominated by diffusion[3,4]. The upper generations of branches are designed to slow down the gas flow to a rate compatible with the rate in the diffusional regime (Pé ~ 1)[5], resulting in uniform distribution of entropy production in both regimes[3,4].

By employing a three-dimensional (3D) fractal structure as flow field inlet channel, we aim to yield similar benefits from replicating these characteristics of the human lung. The fractal pattern consists of repeating "H" shapes where daughter "H's" are located at the four tips of the parent "H". The fractal geometry obeys Murray's law, much like the human lung, hereby leading to minimal mechanical energy losses. Furthermore, the three-dimensional branching structure provide uniform local conditions on the surface of the catalyst layer as only the outlets of the fractal inlet channel are exposed to the MEA.

Numerical simulations were conducted to determine the number of generations required to achieve uniform reactant distribution and minimal entropy production. The results reveal that the ideal number of generations for minimum entropy production lies between $N = 5$ and 7[6]. Guided by the simulation results, three flow-fields with $N = 3$, 4 and 5 (10 cm^2 surface area) were 3D printed via direct metal laser sintering (DMLS), and experimentally validated against conventional serpentine flow fields. The fractal flow fields with $N = 4$ and 5 generations showed ~20% and ~30% increase in performance and maximum power density over serpentine flow fields above 0.8 A cm^{-2} at 50% RH[6]. At fully humidified conditions, though, the performance of fractal $N = 5$ flow-field significantly deteriorates. The susceptibility towards flooding of the high-generation fractal flow-fields is due to the lack of convective gas flow requisite for effective liquid water removal, resulting in significant water accumulation in the interdigitated outlet channels. This was recently confirmed using neutron radiography in an *in situ* water visualisation study.

Another defining characteristic of the fractal approach is scalability, which is an important feature in nature. Fractal flow-fields can bridge multiple length scales by adding further generations, while preserving the building units and microscopic function of the system. Larger, 3D printed fractal flow-fields (25 cm^2 surface area) with $N = 4$ are compared to conventional serpentine flow-field based PEMFCs. Performance results show that fractal and serpentine flow-field based PEMFCs have similar polarization curves, which is attributed to the significantly higher pressure drop (~ 25 kPa) of large serpentine flow fields compared to fractal flow-fields. However, such excessive pressure drop renders the use of a large scale serpentine flow field prohibitive, thus favouring the fractal flow-field.

Lastly, a novel water management strategy that draws inspiration from the passive directional liquid water transport across the skin of the thorny devil is presented. The thorny devil is an Australian lizard with ridged scales, which collects water from any part of its body by simply touching water. The water is transported to the mouth across the skin via capillary action. The phenomenon is based on geometric principles, namely on a periodic pattern of interconnected capillary channels[7]. Using a one-step processing technique, we reproduced the passive water transport phenomenon through fabrication of interconnected capillary channels on the surface of graphite flow field. Preliminary results indicate more than a twofold improvement in fuel cell power density with the integration of capillary channels onto the surface of a parallel flow field, due to superior liquid water separation and removal. Implementation of this water management strategy could circumvent remaining problems of high-generation fractal flow fields.

References

[1] Wu, H.-W., A review of recent development: Transport and performance modeling of PEM fuel cells. Applied Energy, **2016**. 165: p. 81-106.

[2] Hsieh, S.-S., B.-S. Her, and Y.-J. Huang, Effect of pressure drop in different flow fields on water accumulation and current distribution for a micro PEM fuel cell. Energy Conversion and Management, **2011**. 52(2): p. 975-982.

[3] Kjelstrup, S., Coppens, M.-O., Pharoah, J., and Pfeifer, P., Nature-inspired energy-and material-efficient design of a polymer electrolyte membrane fuel cell. Energy & Fuels, **2010**. 24(9): p. 5097-5108.

[4] Trogadas, P., Ramani, V., Strasser, P., Fuller, T.F., and Coppens, M.-O., Hierarchically Structured Nanomaterials for Electrochemical Energy Conversion. Angewandte Chemie International Edition, **2016**. 55(1): p. 122-148.

[5] Coppens, M.-O., A nature-inspired approach to reactor and catalysis engineering. Current Opinion in Chemical Engineering, **2012**. 1(3): p. 281-289.

[6] Trogadas, P., Cho, J.I.S., Neville, T.P., Marquis, J., Wu, B., Brett, D.J.L., and Coppens, M.-O., A lung-inspired approach to scalable and robust fuel cell design. Energy & Environmental Science, **2018**. 11(1): p. 136-143.

[7] Comanns, P., Effertz, C., Hischen, F., Staudt, K., Bohme, W., Baumgartner, Directional, passive liquid transport: the Texas horned lizard as a model for a biomimetic 'liquid diode'. Journal of The Royal Society Interface, **2015**. 12(109).

537099 Electrospun Particle/Polymer Fiber Mats As Hydrogen/Air Fuel Cell Electrodes

Abstract:

Electrospinning is gaining popularity as a convenient and robust technique for fabricating non-woven mats with sub-micron diameter polymer or particle/polymer fibers. Such mats, with a high activity oxygen reduction reaction catalyst powder and a suitable ionomer binder, can be used as the cathode in proton exchange membrane fuel cells. These nanofiber electrodes provide high interfacial electrode area and sufficient macro and micro porosity for product water removal which are of prime importance. In this talk, a review of recent experimental work on nanofiber mat compositions for hydrogen/air fuel cell cathodes will be presented, where the fibers contain carbon-supported Pt or Pt-alloy powders with perfluorosulfonic acid-based polymer binders. The power output using such a cathode is very high for ultra-low Pt loading. For example, a hydrogen/air fuel cell membrane-electrode-assembly containing a nanofiber TKK PtNi/C catalyst powder cathode and a Pt/C nanofiber anode (where the electrodes were 0.1 mg_{Pt}/cm^2, with a blended binder of Nafion® perfluorosulfonic acid and poly(acrylic acid)) generated 756 mW/cm^2 at 0.65 V, 80°C, 100% RH, and 1 atm back pressure. Procedures for fabricating high particle-loaded nanofibers containing platinum-based catalysts and different ionomer binders will be presented. The effects of catalyst type and binder composition on cathode mass activity, electrochemical surface area, cathode durability, and fuel cell power output at high and low relative humidity conditions will be discussed.

519364 Electrocatalytic Hydrogenation of Biogenic Compounds: Reaction Networks and Mechanisms

Abstract:

Increasing availability of sustainable electricity and its intermittent character is stimulating the interest in storing electrical energy into chemical bonds. In this work, we aim to couple electrolytic H_2 production with chemical transformations of organic compounds aiming valuable chemicals and fuels. Thus, a series of carbon-supported metals have been tested for the electrocatalytic reduction of phenolic compounds, ethers, and aldehydes in ideal aqueous conditions.

Electrocatalytic hydrogenation (ECH) of series of phenolic compounds, di-aryl ethers and aromatic aldehydes, has been performed in C-supported Pt, Pd, Rh, and Ni as working electrodes (cathodes), whereas a Pt mesh is used as counter electrode (anode) and an Ag/AgCl electrode is used as reference. H_2 evolution (HER) is the prevalent side reaction on all metals, thus we define efficiency as the fraction of electrons used to reduce the organic compounds. Another series of experiments has been performed, where cathodic potential is replaced by a flow of H_2 at atmospheric pressure. This is denoted as thermal catalytic hydrogenation (TCH).

The electrocatalytic hydrogenation of aromatic rings and carbonyl groups readily occur on Pt, Rh, and Ni, whereas Pd is highly efficient for reduction in the reduction of carbonyl groups in aldehydes. The rates of carbonyl hydrogenation are higher than those of the saturation of aromatic rings. The rates of ECH of all tested compounds, as well as HER, increase with increasingly negative potentials. However, the efficiency is a strong function of the metal. Pt and Ni, for instance, strongly favor HER leading to low efficiencies. Other strategies to increase reaction rates at room temperature are increasing the concentration of hydronium ions and pressure.

In the conversion of substituted phenols and diaryl ethers, the dominant reaction pathway is hydrogenation to cyclic alcohols and cycloalkyl ethers. Methoxy and benzyloxy groups undergo C-O bond cleavage via hydrolysis (minor route) and hydrogenolysis, which opens pathways to the formation of hydrocarbons. In stark contrast to the series of phenolic compounds, aromatic and pseudo-aromatic aldehydes and ketones undergo only reduction of the carbonyl groups without exhibiting any hydrogenation of the aromatic ring. For instance, benzaldehyde and furfural are selectively converted to benzyl alcohol and furfuryl alcohol, respectively. Increasing the concentration of hydronium ions enable C-O bond cleavage of the alcohols through a bifunctional (acid-metal) mechanism.

The results show that electrocatalysis drives the desired transformations at mild conditions and allows a better control over the reaction pathways than thermal catalysis. Cathodic potentials leads to high rates through increasing the availability of surface H and by providing alternative pathways with low-energy barriers.

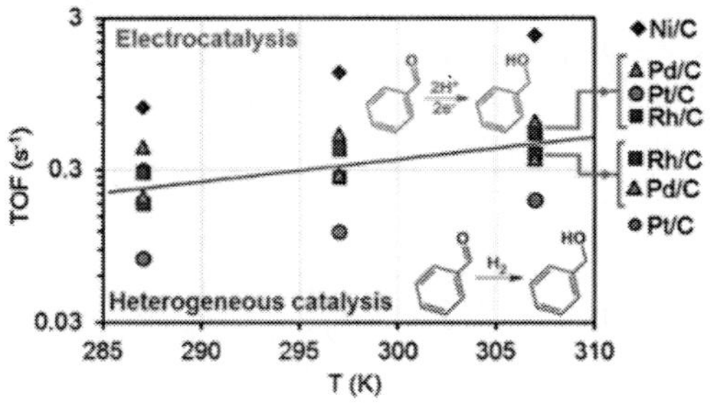

538476 Chemical Transformations Using Electocatalysis: From Small Molecules to Fast Pyrolysis Oils

Abstract:

There is a growing interest in using electrochemical methods for transformation of organic materials. Electrochemical Hydrogenation (ECH) is one such transformation that has multiple applications. Conventional hydrogenation technologies require elevated temperatures and externally supplied high pressure (2000psi) hydrogen. The same reactions can be performed using an electrochemical reactor at close to room temperature and with no externally supplied hydrogen. Hydrogen in ECH is being delivered as a combination of an H^+ and one electron, and the added value comes from avoiding the handling and transport of H_2. ECH has been a focus of interest for niche applications, such as the selective reduction of organic substrates, including biomass conversion. In our studies we have looked three main hypotheses: (1) Selective ECH of the oxygenates occurs through a proton-coupled electron transfer (2) Through tailored electrode structures, ECH can efficiently and economically transform low value and/or low energy materials to high value and high energy products. (3) Mixture components will compete for active catalyst sites and that by controlling the electrode structure and active sites the overall rate can be improved, the desired selectivity increased and the production of molecular hydrogen minimized. Studies were conducted in flow reactors with novel catalysts, using a series of molecules important to biomass conversion and which allow identification of the impact of functional groups on the reaction. We have also studied fast pyrolysis oil upgrading to perform bio-oil hydrogenation to stabilize the more reactive compounds using protons and water as the hydrogen source at room temperature. Promising results have been observed where significant hydrogenation has been found. A description of work conducted and discussion of results will be presented.

531129 Selective Removal of Ethane and Natural Gas Liquids at the Well Pad Via Electrogenerative Processing

Abstract:

Selective removal of ethane and natural gas liquids at the well pad via electrogenerative processing

Maasoomeh Jafari[1], Samgopiraj Velraj[1], and Jason Trembly[1]

1. *Institute for Sustainable Energy and the Environment, Department of Chemical and Biomolecular Engineering, Ohio University, 350 W. State Street, Athens, Ohio 45701, USA*

The U.S. has seen tremendous growth of natural gas liquids (NGLs) supply over recent years, due to the development of unconventional resources. The majority of the new NGL capacity comes from natural gas producing plays (Utica and Marcellus shale) and associated gas from tight oil production (Eagle Ford and Bakken shale). Ethane, the key component of NGLs, is costly and energy intensive to recover from natural gas as cryogenic (turbo-expansion) processing is required. Further ethane storage, handling, and transportation are expensive.

Ethane oversupply has caused the gap between U.S ethane conversion capacity and production. The ethane recovery from NGLs has imposed an additional cost on gas processors. Therefore, the natural gas processors reject ethane to the gas pipeline system to eliminate the recovery cost. However, this technique can not completely manage the ethane overabundance as the gas condensation may occur through transportation under operating pressure. Although the NGLs export is another management technique, most of NGLs plants are far from the coasts.

Cost-effective modular gas separation technologies which may be implemented at the well head are desired. Current cryogenic gas separation technologies will not be cost-effective at individual well head gas throughputs and are not well suited for down scaling. Furthermore, modular methodologies which directly convert NGLs (especially ethane) into more value-added intermediates or materials (such as chemicals or fuels), without need for prior separation, would be the most advantageous as such technologies would reduce flaring of associated gas from oil wells and alleviate mid-stream gas separation bottlenecks.

To address this issue, Ohio University (OHIO) is developing a modular electrogenerative oxidative dehydrogenation (e-ODH) process, which directly converts NGLs at the well head into fungible fuels, electrical power, and pipeline-quality natural gas. In this process, ethane and other NGLs contained in the well head gas are selectively converted into ethylene (alkenes) and byproduct electrical power using a solid oxide fuel cell (SOFC) module, followed by upgrading of the alkenes into gasoline range hydrocarbons and pipeline-quality natural gas in an oligomerization reactor. Advantages offered by this e-ODH process include the following: 1) Modular operation with lower capital and operating costs; 2) Selective conversion of NGLs contained in well head gas; 3) Production of gasoline range hydrocarbons, pipeline-quality natural gas, and electrical power as products; and 4) Utilizes existing SOFC and oligomerization reactor technology minimizing commercial adoption and market entry risk.

To further develop the e-ODH technology, OHIO has been developing Aspen Plus simulations and focusing on the development of high temperature electro catalysts with high ethylene selectivity. Aspen Plus simulation efforts to date have shown the e-ODH process potentially offers well pad operators a novel methodology to convert ethane and associated NGL components into value-added chemicals and byproduct power with an attractive return. This presentation will discuss results from OHIOâ€™s e-ODH Aspen Plus simulation and catalyst development efforts including material characterization and electrochemical performance.

528601 Prediction of the Flash Points of Multicomponent Systems: Applications to Solvent Blends, Gasoline, Diesel, Biofuels and Jet Fuels

Abstract:

Fuel reformulation has been seeded by the consciousness of the potential damages their combustions are causing. Removing one or several components or adding molecules, to reduce pollutants emissions for instance and therefore improve air quality, can modify many properties including safety parameters such as flash points. The flash point of a liquid is the lowest temperature at which a vapor phase at equilibrium with the liquid, is flammable, once mixed with an oxidizer (generally air at atmospheric pressure). For lots of fuel properties, the surrogate fuel blends approach is very often preferred because fuels are generally very complex mixtures containing up to hundreds of components. Predictive methods of flash points for such multicomponent systems are rather scarce in the literature.

In this work, we propose a general approach for accurately predicting the flash points of any kind of fuels or mixtures (pure compounds, solvent mixtures, multicomponent fossile and bio-fuels, chemicals, ...). The predictive method is based on a combination of Catoire-Naudet model [1] with the fully predictive COSMO-SAC dsp activity coefficient model [2]. The main advantage of this method is that experimental flash points of the pure substances are not required, contrary to most other predictive methods for flash point mixtures. Excellent predictions of the flash points of mixtures are obtained for a broad variety of systems. The new model predicts that a small amount of alcohol (methanol, ethanol) in biodiesels can dramatically decrease the flash point of the fuel, in agreement with experimental observations.

538325 Molecular Dynamics Simulation of Quaternary Ammonium Polycation Exchange Membrane Fuel Cell: Nanophase-Segregated Structure and Transport Properties

Abstract:

The nanophase-segregated structure and transport properties of quaternary ammonium grafted polycation membranes are investigated using a molecular dynamics simulation method. For this, we develop a new force field from a reference density functional theory modeling with B3LYP and 6-31G** in order to describe the hydroxide anion. The bond stretching force constant is determined to reproduce the quantum mechanical vibrational frequency. The atomic charges are determined by Mulliken population analysis. Through the annealing procedure, the nanophase-segregated structure is developed as a function of water contents, such as 10 wt % and 20 wt %. The extent of nanophase-segregation is evaluated by the structure factor analysis, which can be compared with the experimental small angle scattering data. Once the equilibrium structures are obtained, we will run long MD simulations to analyze the diffusion of water and hydroxide using the mean-square displacement analysis with an assumption of Gaussian diffusion. Previous processes were modified from past results to minimize potential sources of error and better compare the anion exchange membrane fuel cell with the proton exchange membrane fuel cell.

532136 High-Resolution Differential Phase-Contrast (DPC) X-Ray Imaging for Multiphase Fluid Flow in Three-Dimensional Porous Media

Abstract:

High-Resolution Differential Phase-Contrast (DPC) X-ray Imaging for multiphase fluid flow in three-dimensional porous media

Maha Yusuf[1], George Herring[2], Max Yuen[2], Ching-Wei Chang[2], Yao-Te Cheng[2], Lambertus Hesselink[2,3]

[1] *Department of Chemical Engineering, Stanford University, Stanford, California, 94305*

[2] *Department of Electrical Engineering, Stanford University, Stanford, California, 94305*

3 Department of Applied Physics, Stanford University, Stanford, California, 94305

Abstract

X-ray imaging is a very valuable technique to investigate the inner structure of bulk samples. Some of the most important applications include medical diagnostics, security screening at airports, and materials science[1]. Traditional X-ray imaging uses the differences in X-ray attenuation in different parts of the sample to produce an 'image contrast'. Therefore, the usefulness of conventional X-ray imaging is limited in materials with small density variations due to poor image contrast. To significantly improve the image contrast, phase-shift of X-rays can be recorded instead of attenuation[2-3]. At high energies, X-rays undergo phase shift that are orders of magnitude larger than the attenuation. Thus, detecting phase shifts can be much more sensitive than detecting attenuation. A number of methods have been investigated to detect the phase-shift of X-rays such as interferometric methods, propagation-based methods, and analyzer-based imaging[4-5]. However, most of these methods require a coherent X-ray beam/synchrotron radiation for imaging.

Here, we are working on a grating-based differential phase-contrast (DPC) X-ray imaging system that can be used to retrieve quantitative phase information with polychromatic X-ray sources in a laboratory-based setting[6]. DPC differs from traditional X-ray imaging since it provides three outputs (attenuation, phase contrast, dark-field) instead of one, and uniquely identifies unknown materials with improved image contrast based on the material's absorption, phase, and scattering coefficients. DPC X-ray imaging is a promising

non-destructive technique that is particularly useful for the study of fluid interfaces (e.g. fluid flow in porous media) and interfacial aspects with increased image contrast, investigation of multi-phase fluid dynamics in 3-D porous rocks, analysis of real-time mechanisms, and direct pore-scale imaging. These will improve understanding of the fundamental processes in multi-phase fluid flow, thus contributing to enhanced oil recovery.

References:

1. Sakdinawat, D. Attwood, "Nano-scale X-ray imgaging," Nature Photonics 4, 2010, 840-848.
2. Bech, A. Tapfer, A. Velroyen, A. Yaroshenko, B. Pauwels, J. Hostens, P. Bruyndonckx, A. Sasov, F. Pfeiffer, "In-vivo dark-field and phase-contrast X-ray imaging," Scientific Reports 3, 2013, 3209.
3. A. Nugent, T. E. Gureyev, D. F. Cookson, D. Paganin, Z. Barnea, "Quantitative Phase Imaging Using Hard X Rays," Phys. Rev. Lett., 1996, 77, 2961.
4. D. Auweter, J. Herzen, M. Willner, S. Grandl, K. Scherer, F. Bamberg, M.F. Reiser, F. Pfeiffer, K. Hellerhoff, "X-ray phase-contrast imaging of the breast – advances towards clinical implementation," Br. J. Radiol, 2014.
5. W. Wilkins, T.E. Gureyev, D. Gao, A. Pogany, A.W. Stevenson, "Phase-contrast imaging using polychromatic hard X-rays," Nature 384, 1996, 335-37.
6. Takashima, J. Kim, Y.T. Cheng, M. Yuen, J. Wilde, and L. Hesselink, "Design of X-ray Differential Phase Contrast Imaging System for High Energy and Wide Spectrum Xray Applications," in CLEO, 2014.

521673 Understanding Distillation Curves and Pseudocomponents

Abstract:

Refineries process streams with thousands of components. Refinery streams defeat conventional analytic methods for defining precise chemical compositions of the feeds, intermediates, and products. Conventional approaches use pseudo-components to characterize refinery streams. While long established, few plant engineers understand the assumptions and limitations of pseudo-components.

Data for pseudo-component generation comes from distillation curves generated in specific ASTM methods. In rare cases, non-ASTM methods, can also generate distillation data. Distillation methods are well established. However, plant engineers often do not understand their limitations.

The combined lack of knowledge on both distillation curves and pseudo-component generation often leads to useless and deceptive simulations. Three case studies present common problems with refinery simulations. These include:

- Crude distillation
- Hydrotreating
- Catalytic reforming

Particular focus is placed on the following issues:

- How distillation curves are interpreted, converted and used
- Data accuracy on distillation curves and how that introduces modeling errors
- Data accuracy on sampling and how data that does not mass balance creates errors in simulations

These problems consistently undermine design, operating, and troubleshooting efforts with models. Understanding them, and how to reduce their effect, is critical to effective modeling for plant operations.

533447 Design of Relief and Flare System for Liquefied Natural Gas Plant Using Dynamic Simulation

Abstract:

Process system equipment should be protected from pressure increases of internal fluid pressure in vessels, pumps, compressors, or columns. Overpressure is caused by abnormal operating conditions for examples, exposure to fire, operator error, and heating/cooling medium failure. Relief and flare system (RFS) including pressure safety valve, piping network, knock-out drum, and flare stack provides a protection from overpressure in such upset conditions by discharge of toxic or flammable vapor. Practically, the design of RFS complies industrial code, standard or internal guidelines of company. Those methods consider a number of possible scenarios causing overpressure of equipment and calculate maximum relief load of each scenarios for every unit in the plantwide system. The largest relief load of the various scenarios and units is selected for sizing RFS. However, the result of relief load from these conventional approaches are too conservative and lead to excessive equipment size because they are usually based on the unbalanced heat method or empirical value. Oversizing of RFS not only increases unnecessary expanse but also causes flooding and lifting internals when overpressure occurs in column units. The key point of selection of proper pressure-relieving system is how many possible credits reflected to the examined scenarios. Dynamic simulation takes several credits such as system instrumentation and transient responses of the process, which reduces relief capacity under an overpressure situation. Dynamic simulation is recognized for the calculation method of relief load from American Petroleum Institute code and also utilized in industrial field. However, taking credit by dynamic simulation and reduction of RFS size should be carefully because it can be harmful to process safety. Therefore, a general methodology or procedure of relief load calculation using dynamic simulation is needed for process reliability. In this study, a column operation of natural gas liquefaction process is selected to examine the sizing of relieving devices and a formalized design procedure of RFS is presented from analysis of the dynamic simulation results.

538903 Hydraulic Improvement of a Two-Phase Dehydration Unit for Heavy Crudes through CFD Simulation

Abstract:

This work presents a novel methodology for designing two-phase dehydration/desalting units coupling experimental physical models and CFD simulations in order to improve the hydraulic design of a gun-barrel type equipment. This consisted of a two-phase separation process for heavy crude oils (HCOs) and water (W) with brine, within a framework of an in-situ thermal hydrotreating surface process. A series of experimental tests were performed in a pilot plant using a physical model then results were compared with CFD simulations, which allowed assessing distinct effects on the crude oil emulsion distribution, by means of a two-fluid model. This analysis provided relevant information about the effect of different configurations on the flow distributor behavior, for obtaining an optimal configuration that increases the mass transfer during the desalting process on line from the multiphase flow dispersion analysis.

530464 Modeling and Simulation of 1,3 Butadiene Production Process at Turndown Capacity

Abstract:

1,3-butadiene is required in the manufacture of various materials and chemicals for applications in a wide range of industries such as healthcare, automotive, building and construction and consumer durables [1]. It is mainly used in the production of polybutadiene rubber, styrene-butadiene rubber (SBR), styrene-butadiene latex, nitrile butadiene rubber (NBR), acrylonitrile-butadiene-styrene resin (ABS), chloroprene rubber, etc. Almost all 1,3-butadiene is produced from steam cracking processes. Heavier cracker feedstocks produce more butadiene than the lighter ones. As the cracker feedstock around the globe is trending towards the lighter feedstocks because of the less expansive ethane produced from shale gas, the sustained production of 1,3-butadiene in olefin plants is facing big challenges. This causes not only a decrease in butadiene supply, but also forces various butadiene production facilities to run either at reduced capacities or at turndown capacities. As per the current market trends in North America and Europe, 1,3-butadiene production is expected to continue to decrease for next several years [2].

In this paper, steady state and dynamic models are developed in the Aspen Plus (V10) framework for the production of 1,3-butadiene using N-methyl-2-pyrrolidone (NMP) as a solvent at high turndown ratios of feed rates in an olefin plant. This is a very complex process with significant number of recycle streams, which adds to the difficulty of successfully simulating the process in software. In this process, the mixed feed and the NMP solvent first go through a series of columns where extractive distillation is utilized to separate the butanes, butenes and acetylenes. The vapor product from the extractive distillation goes to another series of distillation columns and is purified to the required purity. The bottom product of the extractive distillation column, which is mainly NMP solvent with traces of hydrocarbons, is purified and recycled to the first and third extractive distillation columns. The maximum operating feasibility to attain on-spec product of the extraction process is examined via parametric analysis in steady-state simulations. The optimal operating status and control strategies under various facility capacities are studied via dynamic simulations.

References

1. 1,3 Butadiene (BD) Market Analysis by Application (Butadiene Rubber, ABS, SBR, SB Latex, NBR, Hexamethylenediamine), Bio-based Opportunities and Segment Forecasts To 2020. (http://www.grandviewresearch.com/industry-analysis/butadiene-market)
2. The Continuing Quest for Butadiene by Jeffrey S. Plotkin. (https://www.acs.org/content/acs/en/pressroom/cutting-edge-chemistry/the-continuing-quest-for-butadiene.html)
3. Butadiene Extraction Technology overview by CB&I.

542630 Suitability of Alternative Aviation Fuels to Modern Aircraft: Impact of Fuel Composition on Liquid Phase Oxidation and Deposit Growth in Aircraft Fuel Systems

Abstract:

The aviation sector has undergone major changes in the recent years to adapt to environmental, economic and energy context. This led, in one hand, to the increasing use of alternative fuels as a means to diversify the energy sources and reduce the environmental impact of aviation and, in the other hand, to the development of high-performance combustion systems which involve more severe thermodynamic conditions. These evolutions were however accompanied by a more frequent occurrence of fuel deterioration and injection system durability issues [1-3]. This poster proposes a study of liquid-phase reactivity of aviation fuels. 21 different conventional and alternative jet fuels were selected. Deposit growth was measured on a heated metal surface as a function of surface temperature from 255 to 380°C using the Jet Fuel Thermal Oxidation Tester associated to thickness measurement by advanced optical interferometry tube rating. Results show a significant impact of fuel composition on the rate of deposit build-up especially the content and the molecular structure of aromatics [4-5]. Additionally, the study proposes an original global kinetic model to represent the growth of deposit function of temperature and residence time. The model allows to represent adequately the growth of deposit in a large temperature range using few number of tests and to rapidly determine the critical breakpoint temperature. These results encourage further developments towards fit-for-purpose formulations of future alternative jet fuels.

[1] US Energy Information Administration," ISBN 9780160933332.

[2] Intergovernmental Panel on Climate Change IPCC, ISBN 978-1-107-65481-5, 2014.

[3] Benini, E., ISBN 978-953-307-611-9, 2011.

[4] Matthew J. DeWitt, Zachary West, et al., Energy & Fuels 28(6):3696–3703, 2014, doi:10.1021/ef500456e.

[5] Arij Ben Amara, Skander Kaoubi et al. «Fuel 173:98–105, 2016, doi:10.1016/j.fuel.2016.01.040.

537467 Recent Advances and New Directions for Optimization of Production SchedulingÂ in Crude-Oil Refineries

Abstract:

Recent Advances and New Directions for Optimization of Production Scheduling in Crude-Oil Refineries

There is an increasing innovation and improvement in production scheduling of crude-oil refineries. Commercial solvers for optimization of industrial problems are progressively robust and efficient; the processing capabilities of computers has evolved tremendously with concurrent processing; techniques and problem-solving strategies are constantly being created and improved. These factors contribute to achieve more efficient production schedule (in operational and economic point of view) and to better integrate wider scale and scope of refining processes. In this way, several studies in the literature addressed techniques: a) to reduce the complexity of the problems by fixing, excluding or relaxing of part of the problems; b) to facilitate the implementation of the problems (programming) by the schedulers or final users in the plant site; c) to reduce the differences between the model and the plant (better prediction) by using closed-loop parameter feedback; d) to access broader and higher quality data, etc. The intention of this work is to cover the recent advances in modeling and solving aspects related to the crude-oil refinery scheduling and give an overview of the new directions in such technology.

Within the modeling aspects, in the last decades, most of works found in the literature for production scheduling optimization in crude-oil refineries have chosen not to use discrete-time modeling due to the combinatorial complexity of the problem. However, considering their selection of continuous-time, the main disadvantages of this approach are the difficulties of execution of tasks by the operators in the plant and the needs of defining previously the points to divide the time horizon. Today, with advances in computer processing, commercials solvers for optimization (e.g., GUROBI), and problem-solving techniques, automated decision using the discrete-time approach has become feasible within reasonable processing time (CPU) and might be used in large-scale problems involving crude-oil processing [1].

In addition to the technical advances mentioned previously, there are new directions in the development of heuristics to be employed in intuitive aspects of the crude-oil refinery scheduling. These methods include a fixation strategy by considering the feedstock storage assignment before a decomposed formulation that integrates the scheduling in a mixed-integer linear (MILP) problem by testing good local and the global optima in a pool of nonlinear (NLP) problems considering, among other techniques, randomized initial point of the nonlinear variables. The MILP construct the time-varying solutions of setups of unit-operations and their connections for the NLPs. The pre-scheduling feed assignment strategy proposed by Kelly et al. [2], which consists of pre-defining the selection of tanks for storage of crude-oils in the refinery, clustering those with similar characteristics by minimizing the bulk crude-oil qualities when different oils are sharing storage tanks. The MILP-NLP decomposition strategy named Phenomenological Decomposition Heuristic (PDH), proposed by Menezes et al. [3] for strategic planning, reduced the complexity of the problem and enable optimization within minutes (< half hour). In this strategy are optimized, iteratively, an MILP model (or *logistics* problem) considering only logic and quantity variables and an NLP model (or *quality* problem) considering only quantity and quality variables, setting the optimized binary variables in the MILP solution.

Kelly *et al.* [1] solved the MILP-NLP scheduling optimization for a time horizon of seven days considering time-periods of two hours. The authors pre-defined feedstock storage assignment [2] to the fixed in the further MILP-NLP scheduling. The modeling of the industrial-scale crude-oil refinery scheduling problem uses the UOPSS (Unit-Operation-Port-State Superstructure) [4] [5]. The problem includes 5 atmospheric distillation units in 9 operational modes and 35 tanks among storage and feeding/charging tanks. The logistic problem (MILP), with approximately 30 thousand continuous and 30 thousand binary variables, 6.5 thousand equations and 80 thousand inequalities (53,802 degrees of freedom) was solved in 128.8 seconds using 8 multi-tasks (threads) in the CPLEX 12.6 solver. The quality problem (NLP), with 102,539 continuous variables, 58,019 equality and 768 inequality constraints (totaling 44,520 degrees of freedom) was solved in 10.3 minutes in the IMPL 'SLP tool using CPLEX 12.6 solver. The MILP-NLP gap between the two solutions was 3.5% after two PDH iterations. However, the authors did not consider the complete network of process-shops and blend-shops beyond the distillation unit.

Despite numerous studies of crude-oil refining scheduling during the last two decades of literature, only one covered the complete refinery network including process units downstream to the distillation unit. Xu et al. [6] included refining operation units such as reforming, cracking, hydrotreating, as well as blending (mixing of products and additives) and fuel storage. However, there is a relatively small number of variables in the continuous-time model for a time horizon of two weeks (<100 binary variables), making it difficult to use for real cases. Franzoi *et al.* [7] included the oil storage and feeding steps, the distillation unit modeled as cascade towers and the refinery set of process-shops. The authors optimized the problem with about 3,000 binary variables and 4,000 continuous variables for a time horizon of 5 days with periods of 4 hours. In addition, they used a strategy of online parameter feedback to cope with uncertainties and to reduce the offsets or inaccuracies over the life-time of the problem as an effective way to close the gap between predictions and productions.

Using, as example, 4-hour as time-step, it pushes forward to a faster and more efficient scheduling, considering this approach in the direction to an online scheduling engine. According to Gupta et al. (2016a), new information may become available over time and should be used as soon as possible, as may occur changes that make the current schedule suboptimal or even infeasible, such as process interruptions, task delays, oscillations in product demand. In this way, the execution of an online scheduling aims to improve the adaptation to uncertainties as well as computing new information within the scheduling re-optimization, which implies in a potential reduction in costs and profit increase for the refinery production. Besides, according to Gupta *et al.* [8] [9], rescheduling should be performed not only when necessary, but regularly, since it considers new information and contribute to further cost reduction. Recent studies involving re-scheduling strategies were developed by Zhang *et al.* [10] for the front-end problem in a refinery, where production is re-scheduled whenever there is a malfunction in a tank (due to mechanical problems); and by Pattison *et al.* [11] who created a mechanism to re-schedule periodically an air separation unit to compute new information regarding prices and demands; or when there are process or market-related disruptions.

In addition, according to Kelly and Zyngier [12], a continuous cycle of improvement is required to provide process feedback and reduce the deviation between model prediction and actual plant values. Thus, using data and information measured with the use of automated systems significantly improves the reliability of the model, especially using techniques of simultaneous parameter estimation and data reconciliation to improve the quality of the information to be updated.

[1] KELLY, J. D.; MENEZES, B. C.; GROSSMANN, I. E.; ENGINEER, F. Crude-oil Blend Scheduling Optimization of an Industrial-Sized Refinery: a Discrete-Time Benchmark. In Foundations of Computer Aided Process Operations, Tucson, United States, 2017a.

[2] KELLY, J. D.; MENEZES, B. C.; GROSSMANN, I. E.; ENGINEER, F. Feedstock Storage Assignment in Process Industry Quality Problems. In Foundations of Computer Aided Process Operations, Tucson, United States, 2017b.

[3] MENEZES, B. C.; KELLY, J. D.; GROSSMANN, I. E. Phenomenological decomposition heuristic for process

design synthesis of oil-refinery Units. Computer Aided Chemical Engineering, v. 37, p. 1877-1882, 2015b.

[4] KELLY, J. D. The Unit-Operation-Stock Superstructure (UOSS) and the Quantity-Logic-Quality Paradigm (QLQP) for Production Scheduling in the Process Industries. In Multidisciplinary International Scheduling Conference Proceedings, New York, United States, v. 327, 2005.

[5] ZYNGIER, D.; KELLY, J. D. UOPSS: A New Paradigm for Modeling Production Planning and Scheduling Systems. In European Symposium in Computer Aided Process Engineering, London, United Kingdom, 2012.

[6] XU, J.; ZHANG, S.; ZHANG, J.; WANG, S.; XU, Q. Simultaneous scheduling of front-end crude transfer and refinery processing. Computers & Chemical Engineering, 2017.

[7] FRANZOI, R.E.; MENEZES, B.C.; KELLY, J.D.; GUT, J.W. Effective scheduling of complex process-shops using on-line parameter feedback in crude-oil refineries, Process Systems Engineering, 2018.

[8] GUPTA, D.; MARAVELIAS, C. T.; WASSICK, J. M. From rescheduling to online scheduling. Chemical Engineering Research and Design, v. 116, p. 83-97, 2016a.

[9] GUPTA, D.; MARAVELIAS, C. T. On deterministic online scheduling: major considerations, paradoxes and remedies. Computers & Chemical Engineering, v. 94, p. 312-330, 2016b.

[10] ZHANG, S.; WANG, S.; XU, Q. A New Reactive Scheduling Approach for Short-Term Crude Oil Operations under Tank Malfunction. Industrial & Engineering Chemistry Research, v. 54, n. 49, p. 12438-12454, 2015.

[11] PATTISON, R. C.; TOURETZKY, C. R.; HARJUNKOSKI, I.; BALDEA, M. Moving horizon closedâ€loop production scheduling using dynamic process models. AIChE Journal, 2016.

[12] KELLY, J. D.; ZYNGIER, D. Unit-operation nonlinear modeling for planning and scheduling applications. Optimization and Engineering, v. 18, p. 133â€"154, 2017.

535905 Refinery-Wide Scheduling for Optimization of Multiple Unit-Operations in the Supply, Production, and Demand Chains in Fuels, Lubes, Asphalts and Petrochemicals Industries

Abstract:

We present a refinery-wide scheduling optimization problem found in every process industry with mixing, transforming and separation where the transforming and separation activities in the crude-oil refinery industry are referred to as reacting/converting processes and distilling/fractionating processes, respectively. For this study we highlight only the quantity and logic aspects or phenomena of the problem (what we term *logistics*) using mixed-integer linear programming (MILP). The quantity and quality details of the problem (what we term *quality*) using nonlinear programming (NLP) are not described. The NPL attributes such as densities, components, properties and conditions require special processing subtypes such as blenders, splitters, separators, reactors, fractionators and black-boxes [1].

In the proposed refinery-wide scheduling approach, the goal is to investigate the limits or capabilities of an enterprise-wide optimization (EWO) by solving simultaneously the supply, production, and demand chains using MILP. This finds time-varying setups of unit-operations and their connections that construct the flowsheet operations to further NLPs by fixing the MILP results. Such EWO setups maybe used in detailed calculations within other edges such as depooling of aggregated tanks by considering their actual topology and operations [2] or decision-automation for production of lubes and asphalts using sequence-dependent switchovers [3].

Figure 1 shows the oil-refinery flowsheet using the unit-operation-port-state superstructure (UOPSS) [4], [5] in a discrete-time modeling. The rectangular shapes ( ) with the cross-hairs are continuous-process types of unit-operations with modes. The triangular shapes are pool unit-operations and the diamond shapes are called perimeters and represent the points where resources or materials enter and leave the flowsheet i.e., sources and sinks. The circles without cross-hairs are inlet ports and with cross-hairs are outlet ports where ports are the flow interfaces to the unit-operations. The lines with arrow-heads (→) connecting outlet-ports to inlet-ports are called external streams or in graph theory they are called arcs or directed-edges. The lines without arrow-heads are internal streams and refer to inlet ports connected to unit-operations and unit-operations connected to outlet ports. The rectangular shapes without cross-hairs are pipeline unit-operations and the upside-down triangular shape is a parcel unit-operation which is modeled as a batch-process and represents the flow of material from one place to another and may represent road, rail and marine modes of transportation [6]. The dotted line boxes surrounding several of the shapes implies that the physical unit has more than one procedural operation associated with it (has multiple modes of operation) and together forms what we call a projectional unit-operation i.e., a physical unit times a procedural operation equals a projectional unit-operation.

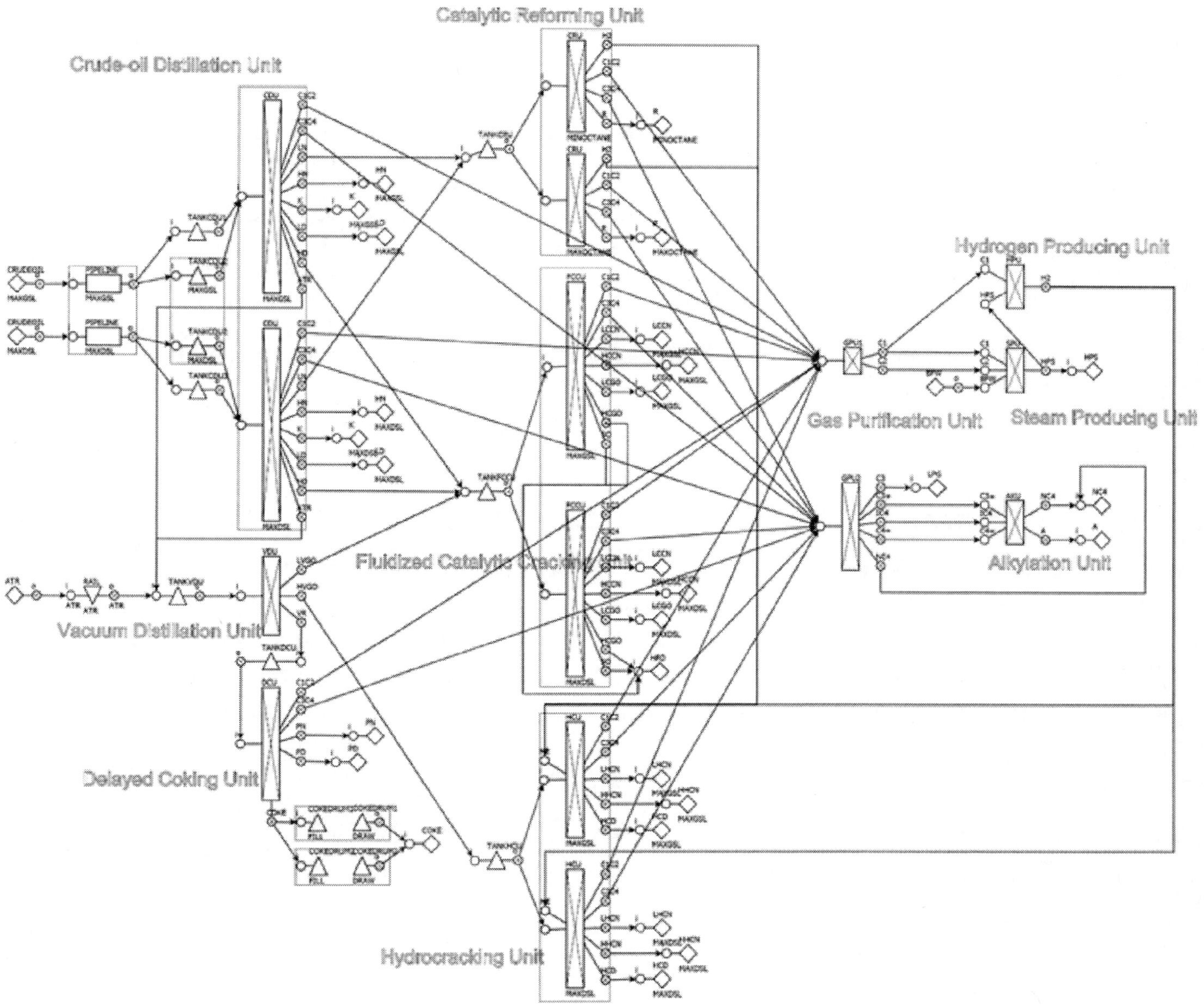

Figure 1. Refinery Process Scheduling Optimization UOPSS Flowsheet.

Each unit-operation and external stream have both a quantity and a logic variable assigned or available and represent either a flow or holdup if quantity and either a setup or startup if logic. Continuous-processes have flows and setups and pools have holdups and setups where perimeters only have a logic setup variable. The internal streams have neither explicit nor independent flow and setup variables given that their flows are uniquely determined by the aggregation of the appropriate external streams and their setups are taken from the setup variables on the unit-operation they are attached to. Although Figure 1 shows an almost complete and integrated oil-refinery excepting the crude-oil and product blending and storing areas, we highlight only the Crude-oil Distillation Unit (CDU), Vacuum Distillation Unit (VDU) and Delayed-Coking Unit (DCU). The other included major process units known as the Catalytic Reforming Unit (CRU), Fluidized Catalytic Cracking Unit (FCCU), Hydrocraking Unit (HCU), Gas Processing Units

(GPU1 and GPU2), Alkylation Unit (AKU), Hydrogen Producing Unit (HPU) and Steam Producing Unit (SPU) are not discussed further.

The two-mode CDU with operations of maximum gasoline (MAXGSL) and maximum diesel (MAXDSL) is of interest given that we have included a pipeline with significant holdup where two segregated crude-oil mixes labeled MAXGSL and MAXDSL are transported from the crude-oil blending and storing area located several kilometers away from the oil-refinery typically at an off-site terminal. Following the pipeline modeling found in Zyngier and Kelly [6] and given 30 one-day discrete time-periods, a pipeline with two modes (MAXGSL and MAXDSL) can be arbitrarily shutdown at any time (i.e., has zero or no flow) in any time-period and when flowing has a constant nominal flowrate. As can be seen in Figure 2 for the Gantt chart resource row "PIPELINE,MAXGSL,o," in time-period 3 or day 3 there is a flow out of the pipeline of the MAXGSL crude-oil mix even though the flow into the pipeline in time-period 3 is MAXDSL i.e., see Gantt chart row "PIPELINE,MAXDSL,i,". In the next time-period (time-period 4) there is a flow of MAXGSL into the pipeline but instead MAXDSL flows out of the pipeline. This is directly related to the fact that a pipeline is modeled as a first-in-first-out (FIFO) queue or renewable resource.

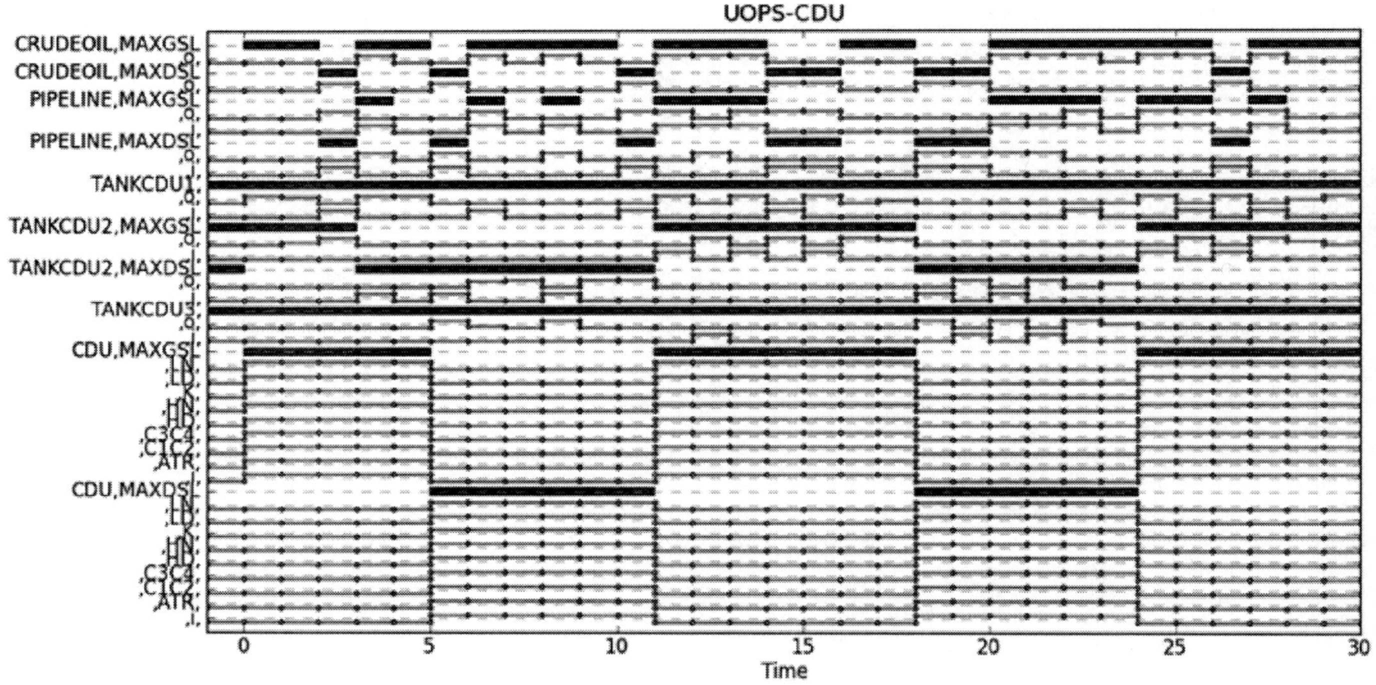

Figure 2. Gantt Chart for CDU with 1-day and 30-day Past and Future Horizons.

There are two other necessary logic features of the CDU's feed operation and they are what we call multi-product and standing-gauge tanks. The multi-product logic is implemented for TANKCDU2 which can store MAXGSL and MAXDSL but not at the same time or in the same time-period. This means that when the optimizer decides to switch the tank's material service from one to the other, the holdup or heel in the tank must be at a certain amount specified by the user and in this case, it is set to zero. The standing-gauge logic, also known as a dead-tank, ensures that if there is flow into the tank then there can be no flow out of the tank and vice-versa. This is configured for the three crude-oil tanks and is apparent from the Gantt chart where the grey trend lines below the black horizontal bars indicates the flow in (port-state "i,") and flow out (port-state "o,"). The standing-gauge can be generalized to restricting that if there is flow in then the flow out be at least some number of time-periods after and this can be used to model mixing- or certification-delays for example.

Figure 3 highlights the operation of the VDU which receives atmospheric residue or reduced crude-oil (ATR) from both the upstream CDU and from an off-site oil-refinery imported by rail. In this situation, the VDU feed ATR is transported by what are known as unit-trains which will usually have around 100 tanker rail-cars per train. There is a round-trip or travel-time in this case of 4-days where on the first day the ATR is loaded, on the second day it is hauled from the source oil-refinery or terminal to the destination oil-refinery, on the third day it is unloaded and on the fourth day it travels back to the source and this is called industrial-shipping as opposed to liner or tramp shipping. On the Gantt chart row labeled "TANKVDU,i," we see the extra amount of ATR being received into TANKVDU from the parcel unit-operation representing the unit-train deliveries of ATR.

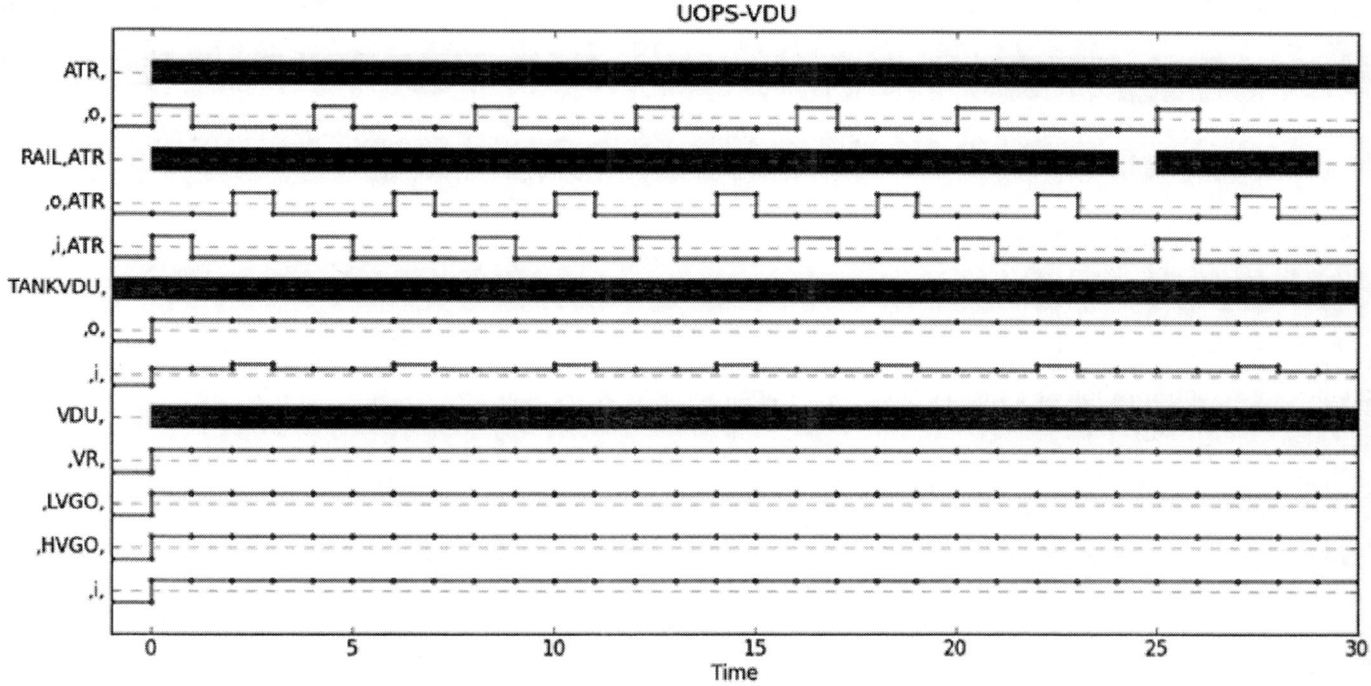

Figure 3. Gantt Chart for VDU with 1-day and 30-day Past and Future Horizons.

Figure 4 displays the Gantt chart for the DCU operation which converts the vacuum residue labeled as VR into pyrolysis naphtha (PN) and distillate (PD) as well as light gases such as methane, ethane (C1C2), propane, propylene, butanes and butylenes (C3C4) including a solid coke material (COKE). The interesting aspect of the DCU is the production of coke which is modeled as a semi-batch arrangement using pool unit-operations. The two physical coke drums COKEDRUM1 and COKEDRUM2 have two operations or modes of FILL and DRAW where the FILL operation can only charge material and the DRAW mode can only discharge material. This means that if COKEDRUM1 is filling then it cannot be drawing which implies that COKEDRUM2 must be drawing. Conversely, if COKEDRUM1 is drawing then it cannot be filling and therefore COKEDRUM2 must be filling. We can observe this behaviour or functionality on the Gantt chart where we see that when COKEDRUM1 is in the FILL mode COKEDRUM2 is in the DRAW operation and vice-versa.

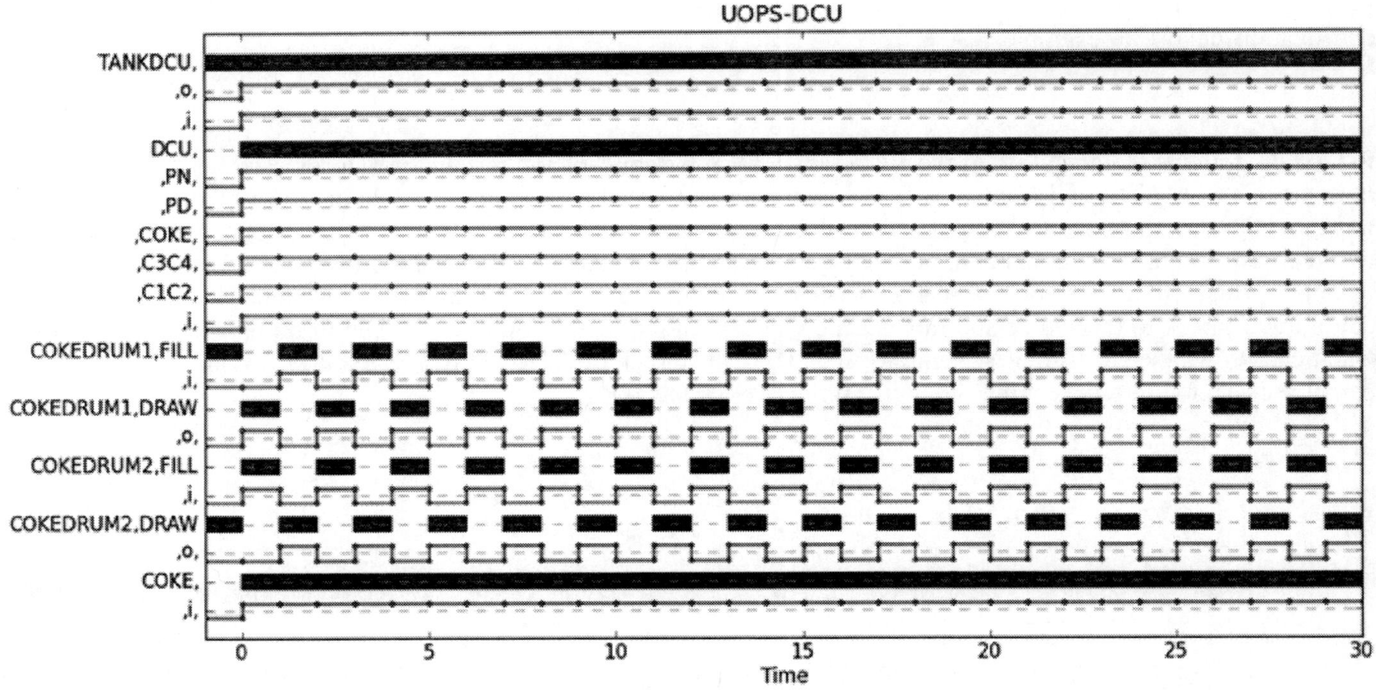

Figure 4. Gantt Chart for DCU with 1-day and 30-day Past and Future Horizons.

[1] Kelly, J. D.; Zyngier, D. "Unit-operation nonlinear modeling for planning and scheduling applications. Optimization and Engineering", v. 18, p. 133–154, 2017.

[2] Menezes, B.C., Kelly, J.D., Grossmann, I.E., "Logistics optimization for dispositions and depooling of distillates in oil-refineries: closing the production scheduling and distribution gap", ESCAPE 2017, June, 2018.

[3] Menezes, B.C., Kelly, J.D., Grossmann, I.E., "Decision Automation for Lubes and Asphalt Production Scheduling using MILP with Sequence-Dependent Switchovers", AIChE Spring Meeting, Orlando, April, 2018.

[4] Kelly, J.D., "The unit-operation-stock superstructure (UOSS) and the quantity-logic-quality paradigm (QLQP) for production scheduling in the process industries", In: MISTA 2005 Conference Proceedings, 327, (2005).

[5] Zyngier, D., Kelly, J.D., "UOPSS: a new paradigm for modeling production planning and scheduling systems", ESCAPE 22, June, 2012.

[6] Zyngier, D., Kelly, J.D., "Multi-product inventory logistics modeling in the process industries", In: W. Chaovalitwonse, K.C. Furman and P.M. Pardalos, Eds., Optimization and Logistics Challenges in the Enterprise", Springer, 61-95, 2009.

540376 The Evaluation of Volatile Char Interaction by the Novel Designed Crucibles in TG

Abstract:

The evaluation of volatile char interaction by the novel designed crucibles in TG

Xiaoming Li (*lxmlily2014@126.com)*

School of Chemical and Biological Engineering,

Taiyuan University of Science and Technology, Taiyuan 030021, PR China

Jin Bai (stone@sxicc.ac.cn)

State Key Laboratory of Coal Conversion, Institute of Coal Chemistry, Chinese Academy of Sciences, Taiyuan 030001, PR China

Abstract:

Coal is still the primary energy in China and it is predicted more coal will be used by gasification instead of combustion. Coal gasification is regarded as the key technology for clean coal utilization. The volatile char interaction during gasification is regarded as one key factor on char structure and char activity. Two novelty crucibles with double layer and separated space are designed to investigate the volatile char interaction during pyrolysis. The crucible A is divided into two layer by the platinum mesh. The volume of upper and bottom of the crucible is equal. This crucible is used for investigate the interaction between volatile and char directly. Compared with crucible A, the bottom part of crucible B is divided into two parts. Although the samples are isolated, but the reaction between volatile from bottom could occur under the mesh. In hence, we can investigate the effect of volatile after reaction on the char. The crucible with the same design but large size is also used for preparing char sample, and the char structure is analyzed by Raman, and the reactivity is analyzed by TG. The volatile char interaction between one typical anthracite, biomass and one low rank coal are investigated in this work. The results will help to understand the degree of char-volatile interaction during gasification and may help to design the gasifier.

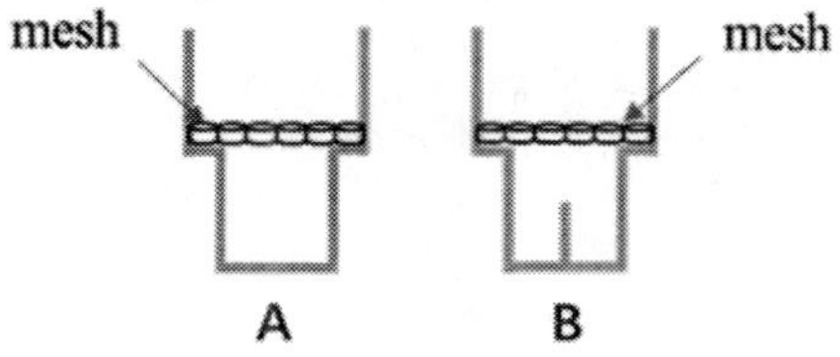

Figure 1. The design of the crucibles for volatile-char interaction by TG

Key words: coal gasification, volatile-char interaction, novel designed crucible, TG

532939 Water Impact of a Gas Shale Production and Distribution System in Mexico

Abstract:

Water Impact of a Gas Shale Production and Distribution System in Mexico

Maria G. Laguna-Martinez[a], Vicente Rico-Ramirez[a] and Jose M. Ponce-Ortega[b]*

[a]Tecnologico Nacional de Mexico en Celaya, Departamento de Ingenieria Quimica, Av. Tecnologico y Garcia Cubas S/N, Celaya, Guanajuato, Mexico 38010

*Corresponding author: vicente@iqcelaya.itc.mx

[b]Universidad Michoacana de San Nicolas de Hidalgo, Departamento de Ingenieria Quimica, Morelia, Michoacan, Mexico, 58060

Abstract

The US Energy Information Administration (EIA) estimated that Mexico has technically recoverable shale gas resources of 545.2 trillion cubic feet; around 343 trillion cubic feet are located in the Burgos Basin, which is connected to the Eagle Ford shale play in Texas. Based on this information, it is fair to consider Mexico at the top world places concerning shale gas resources in its territory (7th largest reserve). However, the exploitation of shale gas in Mexico still requires further study regarding its impacts on environmental (water footprint, in particular), technical and social aspects.

This paper provides a mathematical programming formulation for the optimal planning of a gas shale production and distribution system in Mexico. The model includes an integrated natural gas production and distribution network and the relevant issues related to the optimal management of the available water resources. The analysis considers a time horizon of 15 years divided into 60 time periods. The case study uses real information provided by the Mexican national oil company (PEMEX) database and the geographic information system ESRI ArcGIS. The case study includes 7 existing natural gas processing plants (Arenque, Poza Rica, Matapioche, La Venta, Cactus and Cd Pemex) and 6 potential shale Basins (Chihuahua, Burro-Picachos, Sabinas, Burgos, Tampico-Misantla and Veracruz) recognized in Mexico. The resulting formulation is an MILP involving 12600 binary variables, 85213 continuous variables and 122719 constraints, which we solved through the GAMS-CPLEX framework. Among the results, a water impact assessment is provided for various scenarios, involving either maximum NPV or minimum fresh water consumption.

Keywords: *Shale gas production, shale gas distribution network, mathematical programming, water footprint.*

533741 *Studies on the Oxidative Stability of Mineral Naphthenic Oils Using Commercial Antioxidants*

Abstract:

Mineral naphthenic oil is normally used in transformers, circuit breakers and other electrical equipment for insulating and cooling functions. Due to their usage conditions, it may be subjected to oxidation reactions caused by the presence of oxygen, water and metals, affecting the oil performance. Using the ASTM D2440 method, oxidation studies were carried out on insulating oil, using commercial antioxidants, to assess and compare their thermal oxidative stability, monitored by FTIR analysis, total acid number and sludge content on the oxidized samples. Furthermore, combinations of antioxidants and metal passivators were evaluated to seek the best concentrations that could be used to improve the stability of the insulating naphthenic oil.

529931 Utilization of *Omto Bimo (Kigelia pinnata)* peels As Economic Biobriquette for Community in Cepu District, Indonesia with Sni-01-6235-2000 As Quality Standard

Abstract:

The broad territory and high population of Indonesia have caused some difficulties in energy distribution to fulfill the demand of this country. Energy producing capacity of Indonesia from petroleum refineries have not been able to cover this high energy demand which caused a high deficit in energy sector. Some methods have been attempted by the government to solve this problem through energy conservation and renewable energy development. A renewable energy source that is potential to be developed in Indonesia as an agrarian country is biomass. But only 3.25% of its potential have been fully developed in Indonesia. Cepu is a district in Central Java famous as an oil producing district. Along the streets and houses in Cepu, there are many *omto bimo (Kigellia pinata)* trees planted only to be used as shade trees. The economic fuels that have been widely used by the community in Cepu is kerosene, LPG, and charcoal which has been declining in production. *Omto bimo* have a potential to be utilized as a biomass source because of its high availability in form of a biobriquette. But to this day, there has been no reference about the utilization of *omto bimo* fruit as a biobriquette.

Biobriquette is made through several processes which are material preparation, carbonization, crushing, molding, and drying. From the data obtained through the experiment, 500 gr charcoal of *omto bimo* can produce 15 biobriquettes with the average content of moisture 5.75%, ash 7.76%, volatile matter 3.59%, fixed carbon 82.90%, calorific value 5708.8 kal/gr, compression strength value 47.15 kg/cm^2, drop test value 0.058%, combustion duration 15.602 minutes, and combustion rate 2.26 gr/minute. According to SNI-01-6235-2000 quality standard of charcoal, these testing results have fulfilled the standard specified by *Badan Standarisasi Nasional* (Indonesian National Standardization Board). From economical point of view, biobriquette price is Rp 0.464/kkal which is lower than kerosene and LPG. For usage in economic activities, using *omto bimo* biobriquette as fuel will save up 40% of fuel expense in 1 month compared to using charcoal. It is hoped that *omto bimo* biobriquette can be mass-produced to be used widely by the community as an economic fuel not only in Cepu District but also in Indonesia.

543635 A Technoeconomic Analysis of the Chloralkali Process for Hydrogen Production Using Solar Energy

Abstract:

The economic viability of the photoelectrochemical (PEC) production of hydrogen with chlorine and sodium hydroxide as co-products is studied using H2A modelling methods. This study intends to guide research in the development of novel PEC technology optimized for hydrogen production by identifying the components of the system with the largest contribution to the levelized cost of hydrogen (LCOH). Calculations are based on a conceptual plant designed for the production of 100 kg per day. Final LCOH is given in terms of production cost per kg of hydrogen produced [$/kg]. The PEC cell considered in this study consists of a light absorbing material, a PDMS stabilizing coating, platinum as the reduction catalyst and DSA catalyst for chlorine production. Parameters in the H2A Model with great impact on the LCOH production are solar-to-hydrogen efficiency and the PEC cell life time which were assigned as 6% and 2 year, respectively. PEC cell replacement cost was found to be the greatest contributor to the LCOH. To mitigate for this cost, revenue obtained from the co-products, chlorine and sodium hydroxide, will be allowed to compound interest for 2 years before being invested back into the plant to pay for PEC replacement expenses.

539922 Modification of ZSM-5 zeolite based additive in FCC process for maximizing propylene production

Abstract:

Nowadays, In FCC process, ZSM-5 uses as an efficient additive to increase light olefins yield, in particular propylene. Propylene is one of the most important building blocks for the petrochemicals Industry. The demand for propylene is growing and its market is expanding. Market analysis show that the demand for propylene is growing faster than that for ethylene and the current supply cannot fulfill the demand. This review aims to highlight the recent and the most important modification methods applied for improving ZSM-5 performance in the FCC process to increase the yield of light olefins in general with a focus on maximizing the production of propylene. These methods include the modification of acidity and particle size, synthesis of mesoporous/ hierarchical ZSM-5 structure, phosphorus treatment, introduction of rare-earth metal and incorporation of alkali metals and transition metals.

541307 Catalysis of Calcium on Fe_2O_3/Al_2O_3 Oxygen Carriers in Chemical Looping Combustion

Abstract:

Catalysis of calcium on Fe_2O_3/Al_2O_3 oxygen carriers in chemical looping combustion

ZhiFeng Zhang, Yifei Wang*, Guangsuo Yu, Fuchen Wang

Key Laboratory of Coal Gasification and Energy Chemical Engineering of the Ministry of Education, East China University of Science and Technology, Shanghai 200237, PR China

Chemical looping combustion technology has the function of capturing CO_2 with near zero energy consumption.[1,2] Oxidation and reduction of oxygen carriers (the substances that transfer oxygen during chemical looping combustion) are carried out in two different reactions, namely air reactor and fuel reactor. The oxygen carrier absorbs oxygen in air reactor, and then transfers the lattice oxygen to fuel in fuel reactor, avoiding the direct contact between the air and fuel, so the concentration of CO_2 in the flue gas is very high. In addition, the process belongs to flameless combustion and the reaction temperature is so low that no NO_x pollutants form. Unlike the chemical looping combustion of gaseous fuels, a certain amount of ash and oxygen carriers will participate in the cycle after the combustion of solid fuels, especially coal. Bao et al. [3] studied the interaction between 4 types coal ash and iron base oxygen carrier. The study showed that most ash will reduce the activity of the oxygen carrier and appear sintering and slagging except the ash mainly containing CaO. Zhang et al. [4] studied the chemical looping combustion process of $Fe_2O_3/CaSO_4$ composite oxygen carrier. It was found that the reactive activity of the compound oxygen carrier was stronger than that of single Fe_2O_3 or $CaSO_4$ oxygen carrier. It indicated that there was synergistic effect between Fe_2O_3 and $CaSO_4$ in chemical looping combustion.

Based on coal based chemical looping combustion, the Catalysis of Ca, a typical active mineral in coal, to iron based oxygen carriers is studied in a tubular fluidized bed reactor. The effects of Ca content in oxygen carriers on the reactivity and combustion product release characteristics are compared. It is found that Ca can significantly increase the reactive activity of oxygen carrier. The oxygen carrier with 5% Ca content has the best reactivity, and the further increase of Ca content is not good for CO transformation. Through repeated redox cycles, it is found that the activity of oxygen carriers containing $CaSO_4$ rapidly declined after third cycles. $CaSO_4$ decomposition is the main reason for the decrease of oxygen carrier activity. The oxygen carrier added to $CaCO_3$ shows good stability. The oxygen carrier composition and apparent morphology are characterized by scanning electron microscope (SEM) and X ray diffractometer (XRD). It was found that $CaAl_2O_4$ and $CaAl_4O_7$ are formed after the Ca content is over 5%, which result in oxygen carriers slagging and sintering.

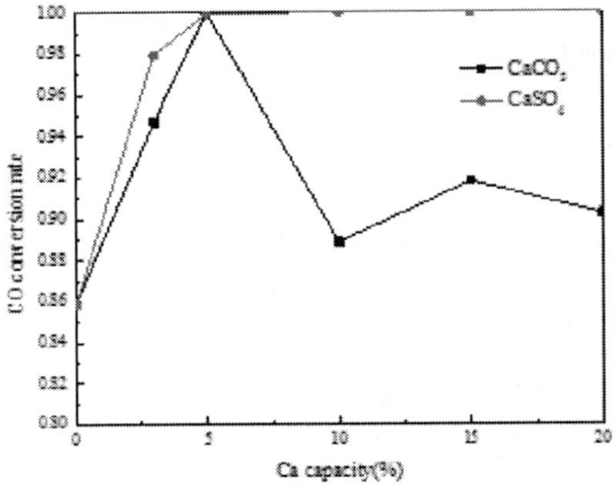

Figure 1. The change of the average conversion rate of CO with the amount of Ca added

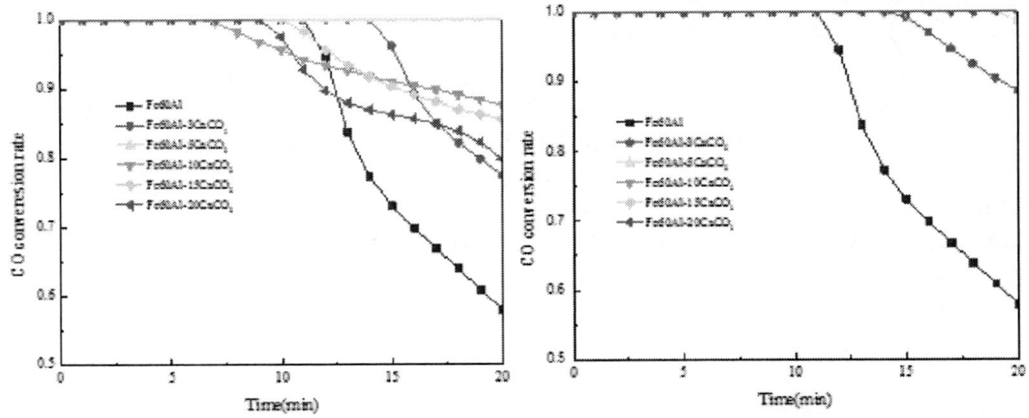

Figure 2. The instantaneous conversion of CO varies with time

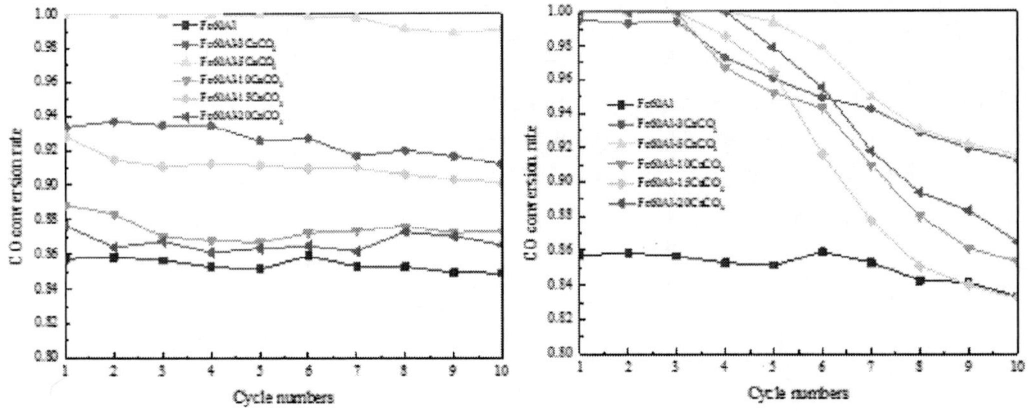

Figure 3 The change of average conversion rate of CO with the number of cycles

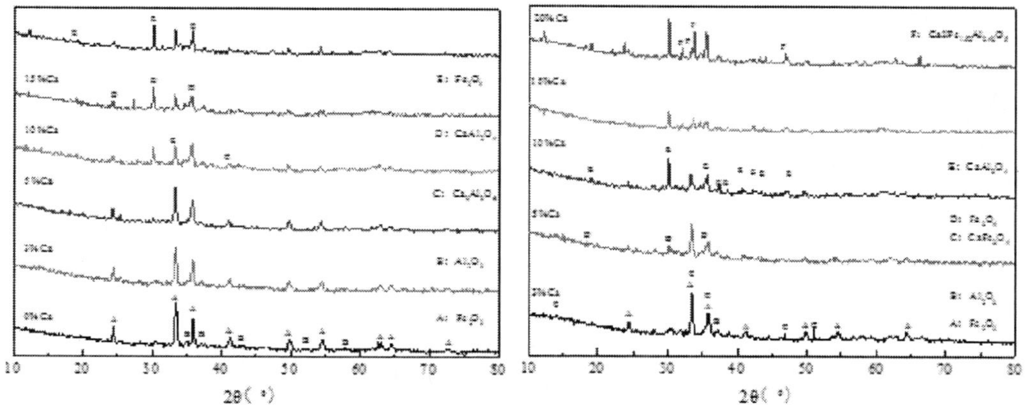

a Add $CaCO_3$ b Add $CaSO_4$

Figure 4 XRD Atlas of oxygen carriers after 10 cycles

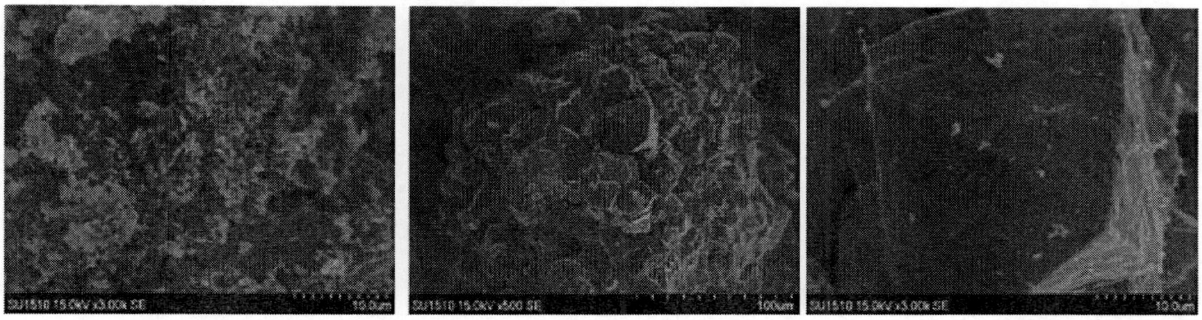

Raw Fe60Al-20$CaCO_3$ Fe60Al-20$CaCO_3$ after 10 cycles

Raw Fe60Al-20$CaSO_4$ Fe60Al-20$CaSO_4$ after 10 cycles

Figure 5 SEM analysis of oxygen carriers

Keywords: chemical looping combustion, Fe_2O_3/Al_2O_3 oxygen carriers, calcium additive, catalysis

[1] Lyngfelt A. 2nd Nordic Minisymposium on Carbon Dioxide Capture and Storage. Göteborg, October 2001.

[2] Moghtaderi B. Review of the recent chemical looping process developments for novel energy and fuel

applications. Energy Fuels 2012, 26 (1), 15-40.

[3] Bao J.; Li Z.; Cai N. Interaction between iron-based oxygen carrier and four coal ashes during chemical looping combustion. Applied Energy 2014, 115, 549-558.

[4] Zhang S, Xiao R, Liu J, Bhattacharya S. Performance of $Fe_2O_3/CaSO_4$ composite oxygen carrier on inhibition of sulfur release in calcium-based chemical looping combustion. International Journal of Greenhouse Gas Control, 2013, 17, 1-12.

---

* Corresponding Author: Tel: +86-21-6425 2522. Fax: +86-21 6425-1312.

E-mail address: wangyf@ecust.edu.cn (Y.F. Wang)

525809 Chemical Looping Partial Oxidation with Dry Reforming (CLPD) of Methane on a Ni-Promoted Fe_2O_3/Al_2O_3

Abstract:

In a chemical looping partial oxidation process, CH_4 is partially oxidized by metal oxide into syngas. The produced H_2/CO ratio is 2, which is desirable for the hydrocarbon production in the Fischer-Trophs reaction. However, since CH_4 also decomposes on the surface of metal, real-time H_2/CO ratio is much higher than 2 and the surface of metal oxide is severely coked. In this talk, CO_2 was utilized as a non-stoichiometric co-feed with CH_4 to derive dry reforming and coke gasification. Ni-enhanced Fe_2O_3/Al_2O_3 was synthesized as an oxygen carrier and a catalyst for dry reforming. Ni was entrapped on the Fe_2O_3/Al_2O_3 as a Ni-aluminate phase on the surface. This can further facilitate dry-reforming and suppress methane decomposition. By incorporating 1 wt% of Ni into the alumina matrix and supplying a nonstoichiometric CH_4-CO_2 mixture feed (CO_2/CH_4 ratio = 0.38), the H_2/CO ratio of about 2 and the CO selectivity of 96.76% were achieved with minimized carbon deposition.

515713 A Solid Mineral (limestone) As a Potential Catalyst for Biodiesel Production from Yellow Oleander Oil (*Thevetia peruviana)*

Abstract:

This work considers the use of a solid mineral (limestone) as heterogeneous catalyst for biodiesel production from yellow oleander seed oil (YOO). Limestone was crushed and reduced to less than 200 μm, the crushed limestone was separated into eight samples, all of which were prepared by calcinating at 700°C, four samples were pre-soaked in methanol while the other four were not. It was noted that pre-soaking pulverized limestone in methanol and calcinating for 5 hours gave the highest calcium content of 58.48% as determined by EDXRF Spectrophotometer (EDX3600B). This was used for biodiesel synthesis. The biodiesel production was performed at catalyst amount of 4 g, reaction time of 70 minutes and methanol/oil ratio of 0.1. The biodiesel yield under this condition was 86.00 % (v/v) this was validated by running three experimental explicate. The physical state of the biodiesel was found to be liquid/reddish-brown at room temperature, acid value of 0.508 (mgKOH/g oil), FFA value of 0.254, iodine value of 73.20 (g I_2/100g oil), saponification value of 26.648 (mgKOH/g oil), cetane number of 234.580 among others. Qualities of the yellow oleander biodiesel (YOB) were compared to standards set by ASTM D6751 and EN 14214.

The effect of reaction time on the yield of biofuel produced from yellow oleander seed oil (Thevetia peruviana)

Daniel T. Oyekunle
Covenant University, Ota, Nigeria.

David O. Oyekunle
Landmark University, Omu-Aran, Nigeria.

Abstract

Biofuel production has continued to gain global increase, the use of biofuel produced from biological resources have resulted in environmentally friendly diesel fuel characterized with release of less harmful gases such as carbon dioxide when compared to fossil diesel. In this study, the effect of reaction time on the yield of biodiesel produced from yellow oleander seed oil was investigated. Trans-esterification procedure was carried out to convert the seed oil to biodiesel, at constant volume of methanol and oil, temperature and catalyst amount of 2 g and 6 g. The result showed that biodiesel yield increases as the reaction time was increased from 50 – 60 minutes, further increase in reaction time resulted in lower yield. It was concluded that reaction time above 60 minutes favored backward reaction.

Keywords: Reaction time, Yellow oleander seed, Biodiesel, Trans-esterification.

Introduction

Biodiesel is a methyl ester produced as an alternative to petro-diesel from renewable biological resources such as vegetable oils (edible and non-edible), animal fats, recycled and waste vegetable oils [1]. Biodiesel is similar to petrol-diesel in combustion properties, it's an environmentally friendly diesel fuel which has continuously received considerable attention over the years. It emits 80% fewer hydrocarbons and 50% less particles, resulting into lower release of greenhouse gas emissions [1]. Biodiesel production from oilseed usually require oil extraction from the seeds. The oil can be extracted by different methods which include; solvent extraction, mechanical extraction, microwave assisted extraction, cold solvent extraction and soxhlet extraction methods. The extracted seed oil are thereafter converted to biodiesel through trans-esterification process. Prior to trans-esterification reaction, oils with high values of free fatty acids (FFA) are esterified first by liquid acids, thereafter by base catalyst to improve the yield [2]. Trans-esterification process can be carried out in a homogeneous and heterogeneous system using alcohol, base, or enzyme catalysts. Methanol and ethanol are widely used in the trans-esterification reaction, owing to its low cost [3]. Vegetable oils are trans-esterified to biodiesel because of its high viscosity by reacting it with alcohol in the presence of acidic or basic catalyst to yield ester and glycerol [4]. Studies such as Betiku and Ajala [5] have trans-esterified yellow oleander oil using plantain peels as heterogeneous catalyst. Heterogeneous catalyst have been reported by Kim et al. [6] to be less corrosive, recoverable, and can be reused.

In this study, the effect of reaction time on the yield of biodiesel produced from yellow oleander seed oil was carried out. Pulverized limestone samples calcinated at 700 ^{0}C was used as heterogeneous catalyst. Biodiesel yield was determined at various reaction time of 50, 60, 70, and 80 minutes at catalyst amount of 2 g and 6 g respectively, keeping other conditions such as volume of methanol, oil, and temperature constant.

Materials and methods

Materials

The material used in this research are yellow oleander seeds. The matured oilseeds were collected from Government Secondary School, Ile Ife, Osun state, Nigeria. All reagents used (including n-hexane, methanol, sulphuric acid, diethyl ether and phenolphthalein) were of analytical grade obtained from Chemical Engineering laboratory Landmark University, Omu-Aran, Nigeria.

Experimental methods

The cleaned yellow oleander seeds were sundried in order to remove initial water content, seed coats were removed by hammer. The seeds were oven dried at 60 ^{0}C for 6 h before they were milled into powder by a blender to increase the surface area which in turn increase the oil yield. Soxhlet extraction method (n-hexane as sol-vent) was used during the extraction process. The percentage yield of oil was determined at the end of the process to be 40 % w/w using equation (1)

$$\%\ Oil\ yield\ ({}^{w}/_{w}) = \frac{Weight\ in\ gram\ of\ extracted\ oil}{Weight\ in\ gram\ of\ powder\ sample} \times 100 \quad (1)$$

Heterogeneous catalyst used in this study was prepared according to the method used by Oyekunle [7]. In this study, pulverized limestone was calcinated in a carbolite AAF 1100 furnace at 700 ^{0}C to convert all the impurity present to its corresponding metal oxide. Variables that affect trans-esterification reaction include as temperature, FFA, moisture content, methanol/oil ratio, catalyst amount, and reaction time [8-9]. FFA of the oil used in this study was 0.9024% which was in agreement with reported studies which suggested that it should be less than 1% w/w [10-12]. Complications such as soap-formation would arise if FFA of the oil is more than 1% w/w. The soap would increase the viscosity of reactants thus, reduce biodiesel yield.

Yellow oleander oil (50 ml) was poured into a round bottom flask after which 7 ml of methanol and 2 g of calcinated pulverized limestone samples were added, the mixture was heated at 60 ^{0}C and stirred using constant temperature magnetic stirrer for 50 minutes, the product was separated using separating funnel. The unreacted methanol in the biodiesel has safety risks and can corrode engine components [12] therefore, the biodiesel was washed with warm distilled water (10% by volume) sprayed over the surface of the ester and stirred gently to remove the entrained impurities and glycerol. The percent yield obtained was determined using equation (2).

$$Yield\ (\%) = \frac{Oil\ (wt)}{Biodiesel\ (methyl\ ester)\ (wt)} \quad (2)$$

This procedure was repeated for equal volume of oil, temperature, methanol, and catalyst amount but at reaction time of 50, 60, 70 and 80 minutes and also for catalyst amount of 6 g, at the same reaction times.

Results

Fig. 1 shows the yield of biodiesel based on the different reaction time ranging from 50 – 80 minutes and at different catalyst amount of 2 g and 6 g. The rate of conversion of yellow oleander oil to biodiesel increased with reaction time. It can be seen that biodiesel yield peaks at 60 minutes for both amount of catalyst, thereafter it declines. This shows that further increase in reaction time would lead to decrease in biodiesel yield under the prevailing condition of the reaction. This could be understood from the viewpoint of reversible reaction as stated by Darnoko [13], reaction time greater than 60 minutes may have favored the backward reaction – production of glycerine, which resulted in lower yield of the product.

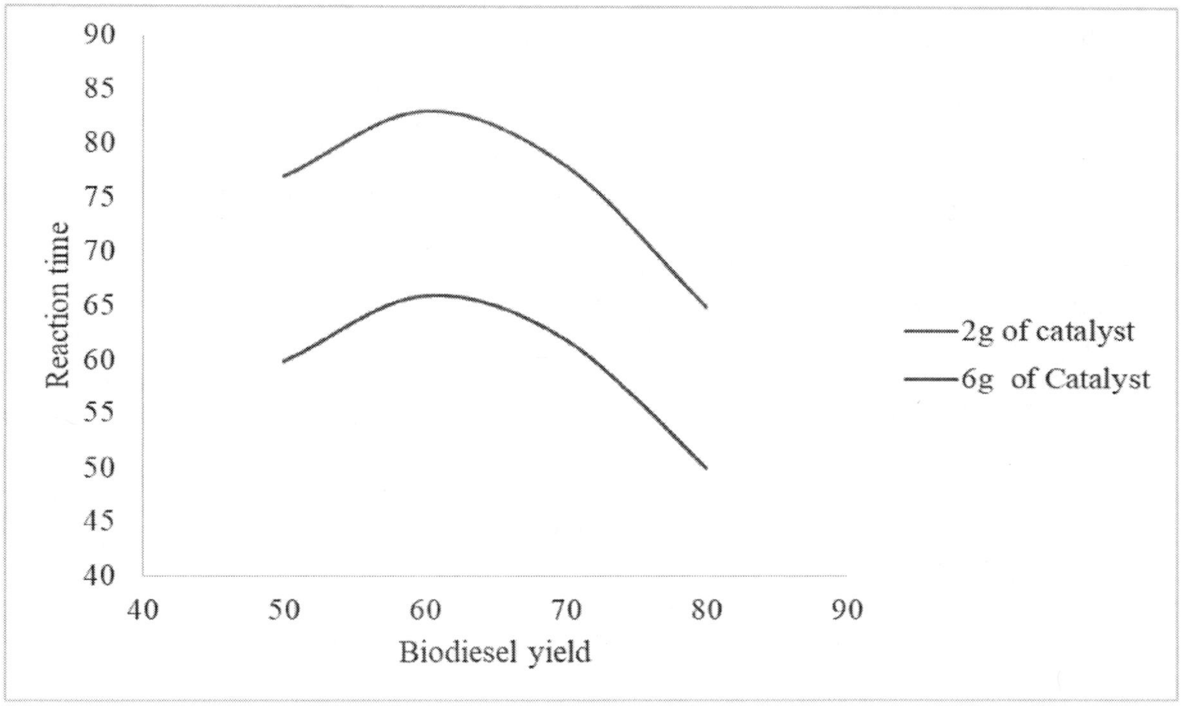

Fig. 1: Effect of reaction time on biodiesel yield for 2g and 6g of catalyst.

Fig 1 illustrate increase in yield as reaction time was increased from 50-60 minutes while every other parameter was kept constant. The yield starts to decrease as the reaction time was increased above 60 minutes for both catalyst amounts. This was because the catalyst has reached equilibrium hence, reverse reaction take place. Higher values of reaction time decrease the biodiesel yield, which will continuously favor the backward reaction. This was in agreement with studies reported by Babagana et al. [14] and Abba et al. [12].

Conclusion

In this study, biodiesel was produced from yellow oleander seed oil and the effect of reaction time (ranging from 50, 60, 70, and 80 minutes) on the yield was carried out. Yellow oleander oil was trans-esterified using calcinated pulverized limestone as catalyst to obtain biodiesel. The result of this study showed that reaction time of 50 – 60 minutes gave an optimum yield of 83 % and 66 % for 2 g and 6 g catalyst used respectively. The result further indicate that lower catalyst amount of 2 g favored higher yield of the biodiesel when compared with catalyst amount of 6 g.

References

1. Wilson P. (2010), Biodiesel Production from Jatropha Curcas. A Review. *Scientific Research and Essays* 5, vol. 14, pp. 1796-1808.
2. Wang Y, Nie J, Zhao M, Ma S, Kuang L, Han X., and Tang S. (2010), Production of Biodiesel from Waste Cooking Oil via a Two-Step Catalyzed Process and Molecular Distillation. *Energy and Fuels*, vol. 24, pp. 2104-2108.
3. Leung DYC, and Guo. Y. (2006) Transesterification of Neat and used Frying oil: Optimization for Biodiesel Production. *Fuel Process Technology*; vol. 87, pp. 883–90.
4. Bello EI, Daniyan IA, Akinola AO and Ogedengbe TI. (2013) Development of a Biodiesel Processor. *Research Journal in Engineering and Applied Sciences*, vol. 2, pp. 182-186.
5. Betiku E, and Ajala SO. (2014), Modeling and optimization of Thevetia peruviana (yellow oleander) oil biodiesel synthensis via Musa paradisiacal (plantain) peels as heterogeneous catalyst: A case of artificial neural network vs. response surface methodology. *Industrial Crops and Products*, vol. 53, pp. 314-322.
6. Kim HJ, Kang BS, Kim MJ, Park YM, Kim DK, Lee JS, Lee KY. (2004), Transesterification of vegetable oil to biodiesel using heterogeneous base catalyst. *Catalysis Today,* vol. 93, pp. 315-320.
7. Oyekunle DT. (2017) A Solid Mineral (Limestone) as a Potential Catalyst for Biodiesel Production from Yellow Oleander Oil (Thevetia peruviana). *CARD International Journal of Engineering and Emerging Scientific Discovery*. vol. 2, pp. 113–127.
8. Meher LC, Dharmagadda VSS, and Naik SN. (2006), Optimization of alkali-catalyzed transesterification of Pongamiapinna-taoil for production of biodiesel. *Bioresources Technology*. vol. 97, pp. 1392-1397.
9. Encinar JM, González JF, Sabio E. and Ramiro MJ. (1999), Preparation and properties of biodiesel from Cynaracarduncu-lus L. oil. *Industrial & Engineering Chemistry Research*. vol. 38, pp. 2927-2931.
10. Tiwari AK, Kumar A, and Raheman H. (2007), Biodiesel production from jatropha oil (Jatropha curcas) with high free fatty acids: An optimized process. *Biomass Bioenergy*, vol. 31, pp. 569-575.
11. Ghadge SV. and Raheman H. (2005), Biodiesel production from mahua (Madhuca indica) oil having high free fatty acids. *Biomass Bioenergy*, 28: 601-605.
12. Abba EC, Nwakuba NR, Obasi SN, Enem JI. (2017), Effect of Re-action Time on the Yield of Biodiesel from Neem Seed Oil. *American Journal of Energy Science*. 4 (2), 5-9.
13. Darnoko D. (1999), Continuous production of methyl esters from oil palm and recovery of beta-carotene by membrane technology, PhD thesis, University of Illinois, Urbana.
14. Babagana G, Shittu SB, Idris MB. (2011) Characterization and Composition of Balanite Aegyptiaca seed oil and its potential as biodiesel feedstock in Nigeria. *Journal of Applied Phytotechnology in Environmental Sanitation*. vol. 1, pp. 29-35.

540959 Liquid Lipase-Catalyzed Esterification of Oleic Acid with Methanol in the Presence of Superabsorbent Polymer for Biodiesel Production

Abstract:

Liquid lipase-catalyzed esterification of fatty acids with methanol is a promising process for biodiesel production. However, water by-product from this process favors the reverse reaction, thus reducing the reaction yield. To address this, superabsorbent polymer (SAP) was used as a water-removal agent in the esterification in this study. SAP significantly enhanced the conversion yield compared with the reaction without SAP. The lipase-catalyzed esterification in the presence of SAP was then optimized by response surface methodology to maximize the reaction conversion. A maximum conversion of 96.73% was obtained at a temperature of 35.25 °C, methanol to oleic acid molar ratio of 3.44:1, SAP loading of 10.55%, and enzyme loading of 11.98%. Under these conditions, the Eversa Transform lipase could only be reused once. This study suggests that the liquid lipase-catalyzed esterification of fatty acids using SAP as a water-removal agent is an efficient process for producing biodiesel.

537395 CO_2 Utilization in a Chemical Looping System for Methane Conversion to High Purity Syngas Using an Iron Based Composite Metal Oxide

Abstract:

Carbon management strategies such as CO_2 utilization to value added products is proving to be a key technology. Chemical looping technology for complete combustion of fuels has been well known in literature for producing CO_2 which is sequestration ready. On the other hand, high purity syngas can be produced by employing a novel co-current moving bed configuration in the chemical looping system which has been well understood on iron-titanium composite metal oxide (ITCMO). Due to ITCMOâ€™s inherent thermodynamic capabilities, it can process CO_2 as a feedstock for methane reforming, thus converting it to an easily disposable product, syngas. This system has been previously investigated from a thermodynamic and process design standpoint, but the reaction mechanism for CO_2 utilization over ITCMO is yet to be explored.

Thermodynamically, the solid phase of $FeTiO_3$ is calculated to be in equilibrium with high purity syngas, thus making this solid phase the primary metal oxide for this study. Experiments on thermogravimetric analyzer (TGA) gave insights on the metal oxide conversion under reduction with methane co-fed with CO_2. Differential fixed bed experiments were also performed to understand the gas phase composition which was measured using a micro-gas chromatography analyzer. Solid characterization techniques such as X-ray diffraction (XRD), X-ray photoelectron spectroscopy (XPS) and Raman spectroscopy were used to understand the changes in the solid phase. These experimental results will help guide the computational analysis in developing the mechanism. Oxygen vacancy interaction mechanism was carefully analyzed using atomistic thermodynamic methods and DFT calculations and compared against the experimental data. The mechanism thus developed will give a better understanding of CO_2 utilization in this novel process, thus providing a pathway to improve the utility of this technology.

530431 Co-Production of Methanol and Dimethyl Ether Using a Zinc Carbonate Modified Catalyst System

Abstract:

A carbonate modified dual-catalytic system of $Cu/ZnCO_3/Al_2O_3$ andÎ³- Al_2O_3 was successfully devised and developed using the SEAMâ€™s proprietary condensed-phase reactive modification technology. This catalyst system was also experimentally investigated in a fully continuous, proof-of-concept mini-pilot scale system for a process engineering study of single-stage syngas-to-dimethylether (ssDME) process. The preliminary experimental results have revealed that (1) the modified catalyst exhibits a good potential to maintain the initial catalytic activity for a prolonged time in a typical reaction environment; (2) the catalyst exhibits high CO_2 tolerance and reactivity for the CO_2-rich feed syngas; and (3) the catalyst possesses an enhanced attrition resistance. This study has also shown that 1) supercritical carbon dioxide is a good medium to convert the conventional $Cu/ZnO/Al_2O_3$ to the modified $Cu/ZnCO_3/Al_2O_3$; 2) the modified catalyst exhibits a reduced CuO crystallite size, and a larger surface area when compared with the original unmodified catalyst; 3) the $ZnCO_3$ modified catalyst provided a better resistance to mechanical attrition during harsh reactor operation; and 4) the modified catalyst shows substantially enhanced anti-leaching characteristics against both water and slurry medium of high-boiling mineral oil. The current concept of pre-reduction modification of the catalyst enables the modified catalyst to be commercially producible as an added benefit.

532820 Experimental Investigation of Combustion Characteristics of a Heating Furnace By Hydrogen/Air Micro-Jet Diffusion Flame

Abstract:

With the depletion of fossil fuel and strict pollution emission regulation in recent decades, the research on the elementary investigation and application of clean alternative fuels has drawn more and more attention [1]. Numerous candidates, such as biofuels, natural gas, ammonia, and hydrogen (H_2), have been proposed as alternative fuels. Among them, Hydrogen (H_2) is considered as one of the most promising candidate with following advantages: (a) H_2 is considered as CO_x and soot emission free when combusted [1]; (b) H_2 are produced from various sources, such as water electrolysis by renewable and nuclear energy [2], fuel reforming from hydrocarbon fuels [3], and gasification from coal and biomass [4]. These merits make H_2-based energy system as an attractive power source, which is important to the safety of the nation energy supply.

The effects H_2 addition on high intensity distributed combustion under swirling flow conditions has been investigated by Khalil and Gupta [5]. The results showed that CO and NO emission were substantially reduced when methane added with H_2. Extremely low emissions of CO (<9 ppm) and NO (<3 ppm) were obtained. Tang et al. comparatively studied the combustion characteristics of methane, propane, and hydrogen fuels in a micro-combustor [6]. They found that H_2/air flame exhibits a wide range of stable flames at various equivalence ratios. The temperature gradient of combustor wall is large because short H_2 micro-jet flames are formed near the inlet, which is different from that of methane and propane flames. The elementary combustion properties have been investigated to apply H_2 as a fuel.

Many researches on combustion characteristics of premixed/non-premixed H_2 added flame, H_2 addition on engine combustion, and fuel cell application have been focused. However, little work has been conducted on the development of H_2 designated burner for thermal application. Therefore, to supply thermal in industrial process, we aim to develop a H_2 designated burner to suit the high burning velocity and short flame length characteristics of H_2 flame. An array type micro-jet burner is proposed and produced in this research. A stainless micro-jet array burner for H_2 combustion is manufactured. The effects of H_2 input power and equivalence ratio on the flame structure, OH radical distribution and furnace temperature of H_2/air diffusion flame on a micro-jet array burner are experimentally investigated. The results show that furnace temperature increased with increasing input power because of intensive formation of OH radicals. The furnace temperature distribution decreased with increasing distance from outlet of the burner and air ratio.

Reference

[1] Alazemi J, Andrews J. Automotive hydrogen fueling stations: An international review. Renew Sustainable Energy Rev 2015;48:483-99.

[2] Huang XN, Gao T, Pan XL, Wei D, Lv CJ, Qin LS, Huang YX. A review: Feasibility of hydrogen generation from the reaction between aluminium and water for fuel cell application. J Power Sour 2013;229:133-40.

[3] Sharma YC, Kumar A, Prasad R, Upadhyay SN. Ethanol steam reforming for hydrogen production: Latest and effective catalyst modification strategies to carbonaceous deactivation. Renew Sustainable Energy Rev 2017;74:89-103.

[4] Wang FC, Zhou ZJ, Dai ZH, Gong X, Yu GS, Wang YF, Yu ZH. Development and demonstration plant operation of an opposed multi-burner coal-water slurry gasification technology. Front Energy Power Eng China 2007;1:251â€"8.

[5] Khalil AEE, Gupta AK. Hydrogen addition on high intensity distributed combustion. Appl Energy 2013;140:71-8.

[6] Tang A, Xu Y, Shan C, Pan J, Liu Y. A comparative study on combustion characteristics of methane, propane and hydrogen fuels in a micro-combustor. Int J Hydrogen Energy 2015;40: 16587â€"16596.

530492 Development of Integrated Geothermal District Heating and Cooling (GDHC) System at West Virginia University Campus-Morgantown, WV

Abstract:

Abstract:

The Morgantown campus of West Virginia University (WVU) is uniquely positioned to host the first geothermal deep direct-use district heating system in the eastern United States. While much of the eastern United States is not blessed with extremely high heat temperatures, the northeastern part of West Virginia is unique in having a basin that is expected to support the achievable flowrate of geofluid through target formations, with sufficient temperatures at depth. These two factors were identified in the 2006 MIT Future of Geothermal Energy Report to be the two most critical factors in minimizing cost of geothermal energy. Our overall project objective is to complete an optimized design for the geothermal system, minimizing the delivered Levelized Cost of Heat (LCOH). In this work, the year-round thermal energy demand on the campus is estimated by recording the steam temperature, pressure and flow rate at steam distribution points. Based on the estimated energy demand and feasible geothermal production flow rate and temperature, different scenarios for surface plant facility are designed. The campus steam demand will be supplied by a geothermal hybrid system, where condensate is preheated using geothermal fluid and it is further heated to required conditions by either natural gas or biomass boilers. Performance of the hybrid geothermal system is modeled using Aspen Plus simulation software. An economic analysis for the geothermal district heating and cooling (GDHC) system is performed using GEOPHIRES (GEOthermal energy for the Production of Heat and electricity Economically Simulated) and feasibility of the GDHC system is determined by comparing costs and benefits with the existing heating and cooling system.

Acknowledgment: This material is based upon work supported by the U.S. DOE's Office of EERE under the GTO, under Award Number DE-EE0008105.

533863 Prevention of Thermal Runaway and Heat Propagation in Battery Module using PCM and Microchannel Plate Cooling System during Nail Penetration

Abstract:

The Lithium-ion (Li-ion) battery has widely applied to electric vehicles and hybrid electric vehicles because of its performance such as high energy density and low self-discharging rate. However, thermal runaway such as fire and explosions can be triggered by thermal abuse operation of Li-ion battery. To relieve this situation, efficient battery cooling system is essential for battery thermal management. We investigate the optimization of mini-channel geometry, coolant flow rate and coolant temperature in liquid cooling system of the Li-ion battery module. Case studies utilizing computational fluid dynamics (CFD) are carried out and each variables is optimized using artificial neural network model (ANN). Nail penetration is employed to estimate thermal management of the proposed liquid cooling system. The temperature distribution, the maximum temperature and thermal runaway propagation are analysed and the results are compared with the other mini-channel cooling systems.

527020 Combustion of an Aqueous Urea/Ammonium Nitrate Alternative Fuel at High Pressure

Abstract:

Combustion of an aqueous urea/ammonium nitrate alternative fuel at high pressure

Bar Mosevitzky, Michael Epstein, Gennady E. Shter, Gideon S. Grader*

*Corresponding author: grader@technion.ac.il

Wolfson Department of Chemical Engineering, Nancy and Stephen Grand Technion Energy Program, Technion-Israel Institute of Technology, Haifa 3200003, Israel

Implementation of renewable energies is hindered by their intermittent power generation profiles. To achieve a truly sustainable energy sector, renewable energy storage is necessary. Chemical fuels are an attractive storage medium since they possess high energy densities. Aqueous solutions of urea and ammonium nitrate (UAN) can act as low-carbon alternative fuels suitable for energy storage applications. Both urea and ammonium nitrate can be synthesized using renewable hydrogen, atmospheric nitrogen and recovered carbon dioxide. Safe to store and transport, UAN combusts under pressure to produce mainly water, nitrogen and some carbon dioxide. However, low concentrations of CO, NO_x and NH_3 were previously detected in UAN combustion effluent. Previous UAN combustion experiments suffered from poorly characterized temperature profiles and complex reactor geometries. Therefore, attempts to simulate the system resulted in poor agreement, impeding unfolding of the chemical reaction network involved in UAN combustion.

In this work, a newly constructed 800-centimeter tubular reactor was used to investigate the combustion of UAN at 1-15 MPa. An inner thermowell tube was used to measure the temperature profile within the reactor (see Figure 1) using a type-K thermocouple, while effluent concentrations were measured using an FTIR spectrometer.

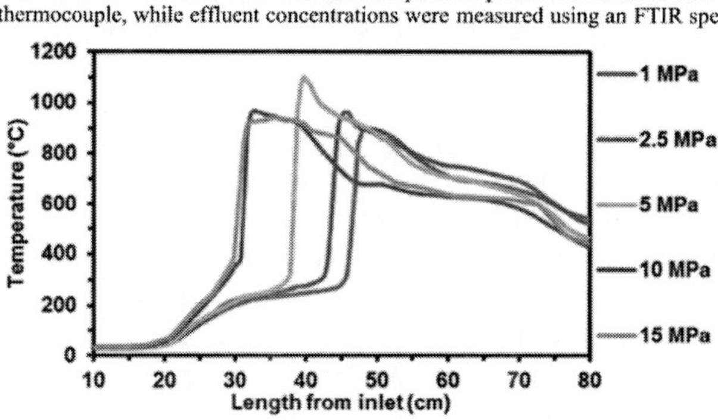

Figure 1. The effect of pressure on the inner UAN combustion temperature profile at 10 ml min^{-1} fuel flow rate.

Theoretical gas-phase calculations were applied to characterize the extent of dispersion of the combustion gasses in the reactor, resulting in values of 0.012-0.028, indicative of near plug flow behavior. Using a PFR model and a newly updated reaction mechanism, the effect of pressure on the effluent composition was simulated and compared to experimental results. Excellent agreement was achieved for all the monitored effluent species, providing the basis for rate-of-production and sensitivity analyses of the reaction pathways. Overall, increasing the system pressure resulted in reduced pollutant emissions and higher N_2 and CO_2 yields (see Figure 2). These simulations indicated that the reaction network involved in UAN combustion becomes less convoluted as the pressure increases, resulting in lower pollutant emissions. Furthermore, the sensitivity of the pollutant levels to the mechanism decreased with pressure, affirming the robustness of the mechanism at high-pressure conditions.

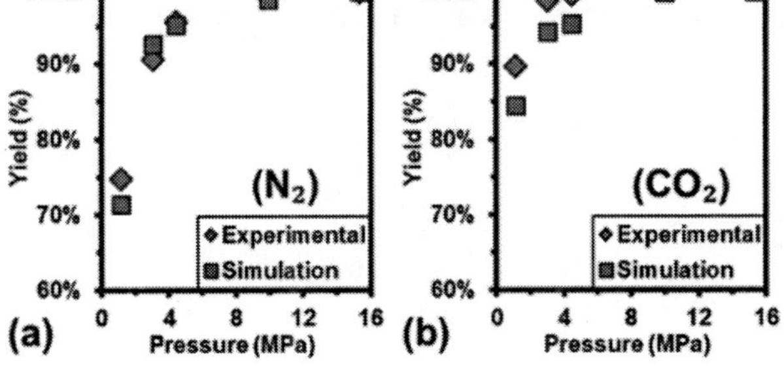

Figure 2. The effect of pressure on the yield of: (a) N_2 and (b) CO_2, as indicated by experimental (♦) and simulation (■) results.

The effect of the combustion pressure of UAN on the temperature profile (Fig. 1), effluent composition (Fig. 2) and the resulting rate-of-production and sensitivity analyses will be presented and discussed.

526251 Reformulation of Gasoline to Replace Aromatics By Biomass-Derived Oxygenates

Abstract:

Reformulation of Gasoline To Replace Aromatics by Biomass-Derived Oxygenates

Gourav Shrivastav[1], Tuhin S. Khan[1], Manish Agarwal[2], and M. Ali Haider[1,*]

[1]*Renewable Energy and Chemicals Laboratory, Department of Chemical Engineering, Indian Institute of Technology Delhi, New Delhi, India*

[2]*Computer Services Centre, Indian Institute of Technology Delhi, India*

Abstract: Gasoline, a complex hybrid of various hydrocarbon families, is an economical, efficient, and ubiquitous fuel. Despite such properties, the major problem with gasoline is its adverse effects on the environment at both local and global scales. This is primarily due to the presence of aromatic components which are added to maintain the octane rating of the gasoline fuel. These compounds are well-known to exhibit toxic and carcinogenic behavior. For decades, methyl tertiary butyl ether (MTBE), an oxygenated additive, has been widely used to improve octane ratings of gasoline. However, now it is proven to be carcinogenic and a major contaminant of groundwater, owing to its high aqueous miscibility. Hence, an environmentally benign oxygenated additive is required for the reformulation of gasoline. In the search for a “green gasoline”, a new reformulation strategy, having no or reduced amount of aromatics, is proposed. Biomass-derived alkyl levulinates are prospected as oxygenated additives as well as blending components to circumvent the use of aromatics and MTBE in gasoline. By utilizing molecular dynamics simulations, the thermophysical and dynamical behavior of gasoline blends with four alkyl levulinates, viz. methyl levulinate, ethyl levulinate, propyl levulinate and butyl levulinate was scrutinized and compared with those of MTBE–gasoline mixtures. It is shown that, at 300 K and 1 atm, ALs in conventional gasoline can be used for reformulation with amounts up to 18 mol% while maintaining the density, viscosity, and compressibility within the recommended limits. However, this amount can be further increased to 35 mol% by modification of aromatic content. Among the studied oxygenates, BL was observed to have the lowest miscibility in water as compared to other ALs studied. The methodology may be applied to study similar biomass-derived oxygenates for their applicability as a fuel additive or blend. In addition, a wide variety of oxygenates, such as alcohol, ketones, esters, diesters, and cyclic ethers, were also scrutinized. The preliminary results for these oxygenates suggest that they can maintain thermophysical and dynamic properties of reformulated fuel but will reduce the energy density.

18AIChE
Annual Meeting, Pittsburgh, PA

H_2 Production via Ferrite based H_2O Splitting Cycle: Solar Reactor Thermodynamic Efficiency Analysis

(Principal Author)
Rahul Bhosale
Department of Chemical Engineering
College of Engineering
Qatar University, Doha, Qatar
rahul.bhosale@qu.edu.qa

(Co-Authors)
Fares AlMomani
Department of Chemical Engineering
College of Engineering
Qatar University, Doha, Qatar

Prepared for Presentation at
American Institute of Chemical Engineers
2018 Fall Meeting
Pittsburgh, PA, USA
October 28 – November 2, 2018

H_2 Production via Ferrite based H_2O Splitting Cycle: Solar Reactor Thermodynamic Efficiency Analysis

Rahul R. Bhosale*, Fares AlMomani

Department of Chemical Engineering
College of Engineering
Qatar University, Doha, Qatar

Keywords: Ferrite, Thermodynamics, Solar Reactor, Efficiency Analysis

Abstract

This study reports the thermodynamic equilibrium and efficiency analysis of the Ba-ferrite based thermochemical water splitting cycle. Attempts were made to identify the equilibrium composition associated with the thermal reduction of Ba-ferrite using HSC Chemistry software. In addition, the equations associated to the thermodynamic efficiency analysis were developed.

1. Introduction

H_2 is a clean fuel with a very high energy density. Production of H_2 via water splitting reaction is considered as one of the promising ways. Direct thermolysis of water is considered as not efficient due to the requirement of a very high temperature. To resolve the issue associated with the high temperature, researchers are significantly working towards the metal oxide based thermochemical H_2O splitting cycle [1-46].

Among the various material investigated, ferrites are utilized heavily towards H_2O splitting reaction, experimentally. However, the determination of the solar-to-fuel conversion efficiency of ferrite based H2O splitting cycles is not much investigated. Therefore, in this study, attempts were made to perform a thermodynamic efficiency analysis of the Ba-ferrite based solar thermochemical water splitting cycle. The chemical reactions associated with this cycle are presented as follows:

$$BaFe_2O_4 \rightarrow BaFe_2O_{4-\delta} + {}^{\delta}/_{2}\, O_2 \quad (1)$$

$$BaFe_2O_{4-\delta} + \delta H_2O \rightarrow BaFe_2O_4 + \delta H_2 \quad (2)$$

HSC Chemistry software and its thermodynamic database was used to conduct the thermodynamic analysis. This analysis is divided into two section: 1. Equilibrium analysis and 2. Efficiency analysis.

2. Equilibrium Thermodynamic Analysis

The equilibrium compositions related to the thermal reduction of the Ba-ferrite were identified by conducting HSC simulation. The obtained results indicate that the thermal reduction of the Ba-ferrite initiated at around 1150 K, and the complete dissociation was observed at around 2500 K. The Ba-ferrite was dissociated into FeO, Fe_2O_3, BaO, and O_2.

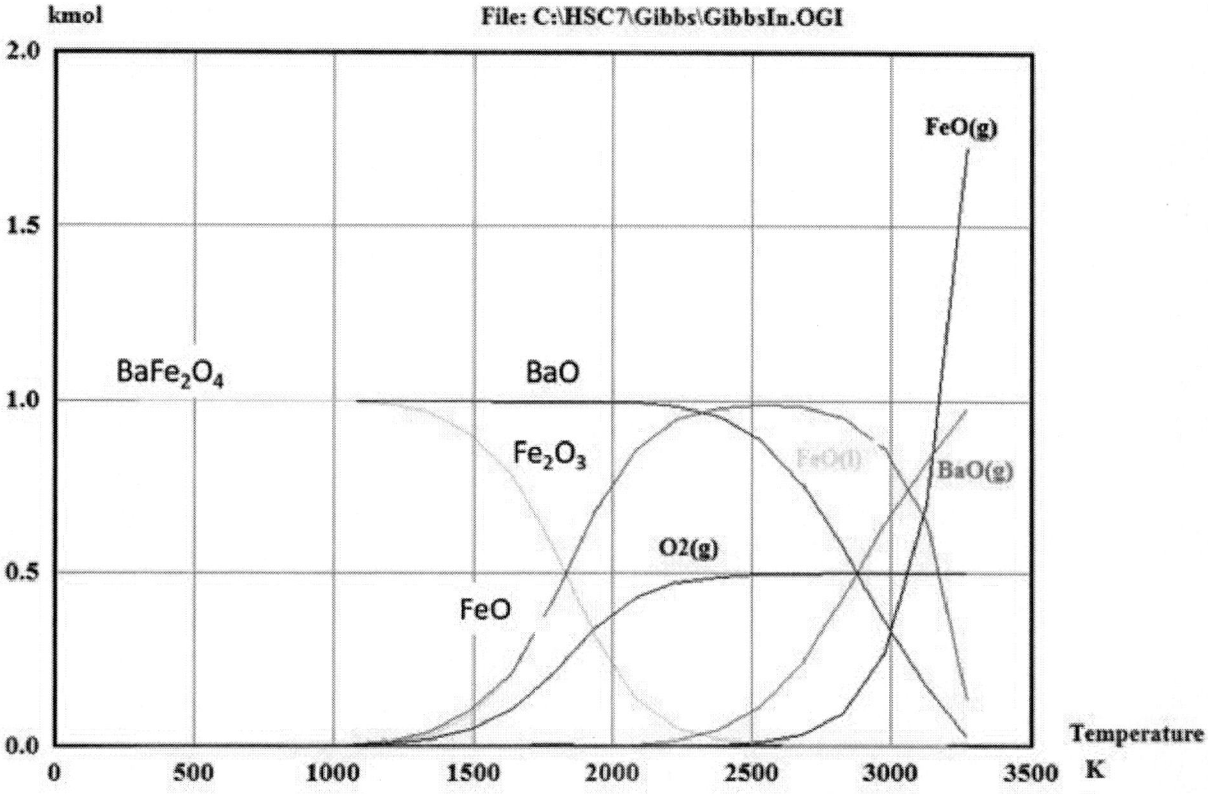

Fig. 1. Equilibrium compositions associated with the thermal reduction of Ba-ferrite.

3. Thermodynamic Efficiency Analysis

The process flow diagram associated with the Ba-ferrite based thermochemical H_2O splitting cycle is shows in Figure 2. The process is consists of a solar reactor and a water splitting reactor, three coolers, one heater and one ideal H_2/O_2 fuel cell. The equations associated with the thermodynamic efficiency analysis are listed on the following page.

The net heat energy required to run the cycles:

$$\dot{Q}_{cycle-net} = \dot{Q}_{BaFe_2O_4-red} + \dot{Q}_{H_2O-heating} \quad (3)$$

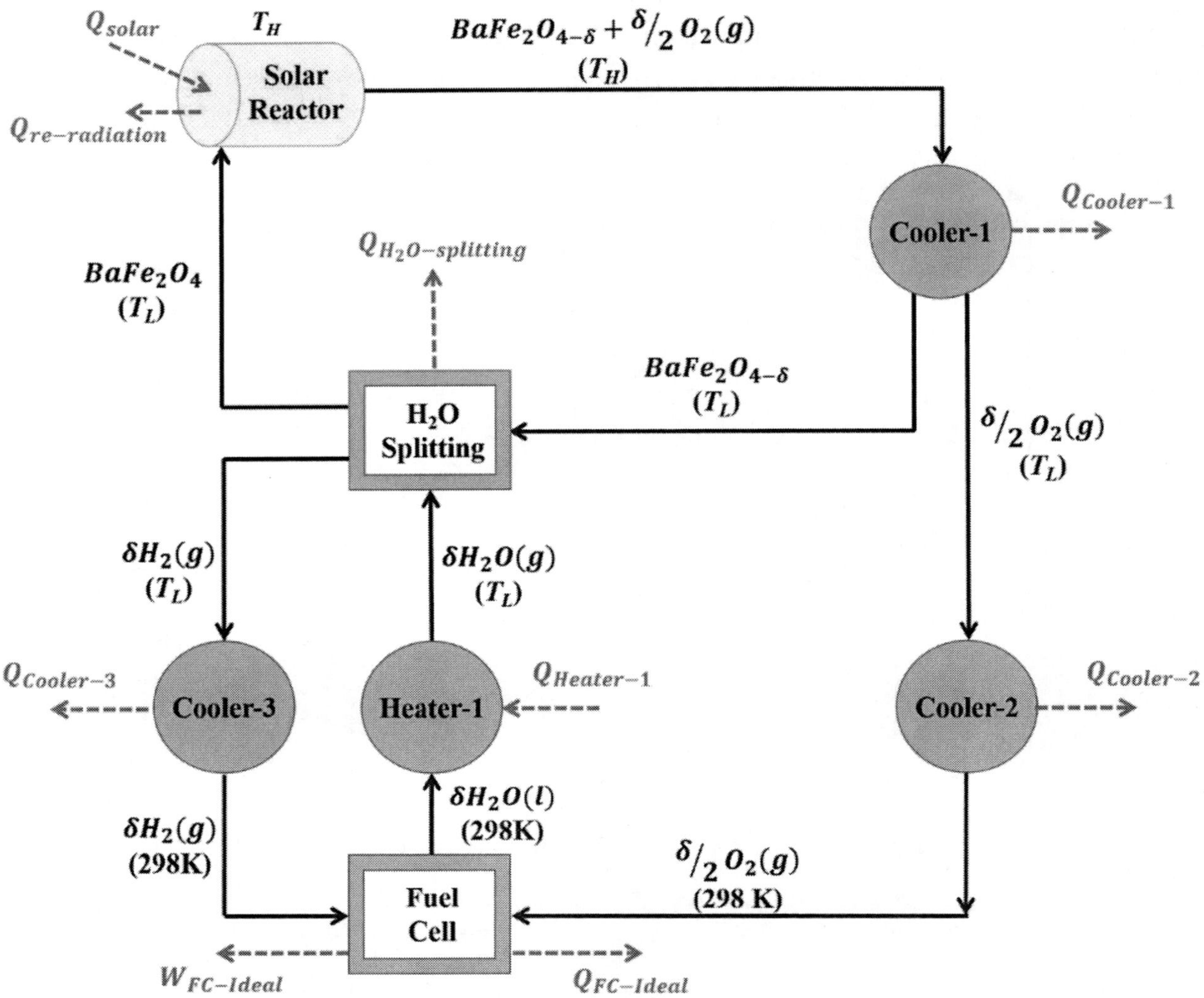

Fig. 2. Process flow diagram associated with the Ba-ferrite based thermochemical H_2O splitting process.

Where,

$$\dot{Q}_{BaFe_2O_4-red} = \dot{n}\Delta H|_{BaFe_2O_4@T_L \rightarrow BaFe_2O_{4-\delta} + \delta/_2 O_2(g)@T_H} \tag{4}$$

$$\dot{Q}_{H_2O-heating} = \dot{n}\Delta H|_{\delta H_2O(l)@298K \rightarrow \delta H_2O(g)@T_L} \tag{5}$$

Total amount of solar energy necessary to drive the cycles:

$$\dot{Q}_{solar-cycle} = \dot{Q}_{solar-reactor} + \dot{Q}_{solar-heater} \tag{6}$$

Where,

$$\dot{Q}_{solar-reactor} = \frac{\dot{Q}_{BaFe_2O_4-red}}{\eta_{abs-solar-reactor}} \quad (7)$$

$$\dot{Q}_{solar-heater} = \frac{\dot{Q}_{H_2O-heating}}{\eta_{abs-solar-heater}} \quad (8)$$

$$\eta_{abs-solar-reactor} = 1 - \left(\frac{\sigma T_H^4}{IC}\right) \quad (9)$$

$$\eta_{abs-solar-heater} = 1 - \left(\frac{\sigma T_L^4}{IC}\right) \quad (10)$$

The re-radiation losses from the cycle:

$$\dot{Q}_{re-rad-cycle} = \dot{Q}_{re-rad-solar-reactor} + \dot{Q}_{re-rad-solar-heater} \quad (11)$$

Where,

$$\dot{Q}_{re-rad-solar-reactor} = \dot{Q}_{solar-reactor} - \dot{Q}_{BaFe_2O_4-red} \quad (12)$$

$$\dot{Q}_{re-rad-solar-heater} = \dot{Q}_{solar-heater} - \dot{Q}_{H_2O-heating} \quad (13)$$

The heat dissipated by the three coolers:

$$\dot{Q}_{cooler-1} = -\dot{n}\Delta H|_{BaFe_2O_{4-\delta} + \delta/_2 O_2(g)@T_H \rightarrow BaFe_2O_{4-\delta} + \delta/_2 O_2(g)@T_L} \quad (14)$$

$$\dot{Q}_{cooler-2} = -\dot{n}\Delta H|_{\delta/_2 O_2(g)@T_L \rightarrow \delta/_2 O_2(g)@298K} \quad (15)$$

$$\dot{Q}_{cooler-3} = -\dot{n}\Delta H|_{H_2(g)@T_L \rightarrow H_2(g)@298K} \quad (16)$$

The heat energy released by the water-splitting reactor.

$$\dot{Q}_{splitting-reactor} = -\dot{n}\Delta H|_{BaFe_2O_{4-\delta} + \delta H_2O(g)@T_L \rightarrow BaFe_2O_4 + \delta H_2(g)@T_L} \quad (17)$$

The highest amount of work that can be obtained and the heat energy that will be liberated by the ideal fuel cell:

$$\dot{W}_{FC-Ideal} = -\dot{n}\Delta G|_{\delta H_2(g) + \delta/_2 O_2(g)@298K \rightarrow \delta H_2O(l)@298K} \quad (18)$$

$$\dot{Q}_{FC-Ideal} = -(298) \times \dot{n}\Delta S|_{\delta H_2(g) + \delta/_2 O_2(g)@298K \rightarrow \delta H_2O(l)@298K} \quad (19)$$

The cycle and solar-to-fuel energy conversion efficiency for the cycle:

$$\eta_{cycle} = \frac{\dot{W}_{FC-Ideal}}{\dot{Q}_{solar-cycle}} \tag{20}$$

$$\eta_{solar-to-fuel} = \frac{\delta \times HHV_{H_2}}{\dot{Q}_{solar-cycle}} \tag{21}$$

3. Conclusions

The equilibrium compositions associated with the thermal reduction of Ba-ferrite were determined and the results indicate that the complete thermal dissociation of the Ba-ferrite is possible at around 2500 K. Furthermore, the process flow diagram and thermodynamics equations associated with the Ba-ferrite based H_2O splitting cycles were developed.

4. Acknowledgments

This publication was made possible by the NPRP grant (NPRP8-370-2–154) from the Qatar National Research Fund (a member of Qatar Foundation). The statements made herein are solely the responsibility of author(s).

References

[1] J.R. Scheffe, A. Steinfeld, Oxygen exchange materials for solar thermochemical splitting of H_2O and CO_2: a review, Materials Today 17 (2014) 341–348.
[2] R. R. Bhosale, G. D. Takalkar, P. N. Sutar, A. Kumar, F. AlMomani, M. Khraisheh, A decade of ceria based solar thermochemical H_2O/CO_2 splitting cycle, International Journal of Hydrogen Energy (2018) Article In-Press.
[3] D. Yadav, R. Banerjee, A review of solar thermochemical processes, Renewable and Sustainable Energy Reviews 54 (2016) 497 – 532.[4] R. R. Bhosale, A. Kumar, F. AlMomani, R. B. Gupta, Solar thermochemical $ZnO/ZnSO_4$ water splitting cycle for hydrogen production, International Journal of Hydrogen Energy 43 (2017) 23474-23483.
[5] P.G. Loutzenhiser, A. Steinfeld, Solar syngas production from CO2 and H2O in a two-step thermochemical cycle via Zn/ZnO redox reactions: thermodynamic cycle analysis, International Journal of Hydrogen Energy 36 (2011) 12141–12147.
[6] R. R. Bhosale, A. Kumar, P. Sutar, Thermodynamic analysis of solar driven SnO_2/SnO based thermochemical water splitting cycle, Energy Conversion & Management 135 (2017) 226-235.
[7] A. Stamatiou, P. Loutzenhiser, A. Steinfeld, Solar syngas production from H_2O and CO_2 via two-step thermochemical cycles based on Zn/ZnO and FeO/Fe_3O_4 redox reactions: Kinetic analysis, Energy & Fuels 24 (2010) 2716-2722.

[8] R.R. Bhosale, A. Kumar, F. AlMomani, Solar Thermochemical Hydrogen Production via Terbium Oxide Based Redox Reactions, International Journal of Photoenergy 2016 (2016) 9727895.
[9] P. Loutzenhiser, M. Gálvez, I. Hischier, A. Stamatiou, A. Frei, A. Steinfeld, CO_2 splitting via two-step solar thermochemical cycles with Zn/ZnO and FeO/Fe_3O_4 redox reactions II: Kinetic analysis, Energy & Fuels 23 (2009) 2832-2839.
[10] R. Bhosale, R. Shende, J. Puszynski, Sol-gel derived $NiFe_2O_4$ modified with ZrO_2 for hydrogen generation from solar thermochemical water-splitting reaction, Materials Research Society Symposium Proceedings 1387 (2011) 34 – 39.
[11] P. Charvin, S. Abanades, F. Lemont, G. Flamant, Experimental study of SnO2/SnO/Sn thermochemical systems for solar production of hydrogen, AIChE Journal 54 (2008) 2759-2767.
[12] R. V. Shende, J. Puszynski, M. Opoku, R. Bhosale, Synthesis of novel ferrite foam material for water-splitting application, Proceedings of the 2009 NSTI Nanotechnology Conference and Expo., 1 (2009) 201 – 204.
[13] C.C. Agrafiotis, C. Pagkoura, A. Zygogianni, G. Karagiannakis, M. Kostoglou, A.G. Konstandopoulos, Hydrogen production via solar-aided water splitting thermochemical cycles: Combustion synthesis and preliminary evaluation of spinel redox-pair materials, International Journal Hydrogen Energy 37 (2012) 8964-8980.
[14] R.R. Bhosale, A. Kumar, F. AlMomani, U. Ghosh, M. S. Anis, K. Kakosimos, R. V. Shende, M. A. Rosen, Solar hydrogen production via a samarium oxide-based thermochemical water splitting cycle, Energies 9 (2016).
[15] M. Roeb, N. Gathmann, M. Neises, C. Sattler, R. Pitz-Paal, Thermodynamic analysis of two-step solar water splitting with mixed iron oxides, International Journal of Energy Research 33 (2009) 893-902.
[16] R. R. Bhosale, A. Kumar, F. AlMomani, U. Ghosh, P. Sutar, G. Takalkar, A. Ashok, I. Alxneit, Effectiveness of Ni incorporation in iron oxide crystal structure towards thermochemical CO_2 splitting reaction, Ceramics International 43 (2017) 5150-5155.
[17] J.R. Scheffe, J. Li, A.W. Weimer, A spinel ferrite/hercynite water-splitting redox cycle, International Journal of Hydrogen Energy 35 (2010) 3333-3340.
[18] R. R. Bhosale, A. Kumar, F. AlMomani, U. Ghosh, D. Dardor, Z. Bouabidi, M. Ali, S. Yousefi, A. AlNouss, M. S. Anis, M. H. Usmani, M. H. Ali, R. S. Azzam, A. Banu, Solar co-production of samarium and syngas via methanothermal reduction of samarium sesquioxide, Energy Conversion and Management 112 (2016) 413 – 422.
[19] J.R. Scheffe, A. Steinfeld, Thermodynamic analysis of cerium-based oxides for solar thermochemical fuel production, Energy & Fuels 26 (2012) 1928-1936.
[20] R. R. Bhosale, R. V. Shende, J. Puszynski, Sol-gel synthesis of ferrite foam materials for H2 generation from water-splitting reaction, Proceedings of the 2010 NSTI Nanotechnology Conference and Expo., 1 (2010) 368 – 371.

[21] W.C. Chueh, C. Falter, M. Abbott, D. Scipio, P. Furler, S.M. Haile, A. Steinfeld, High-flux solar-driven thermochemical dissociation of CO_2 and H_2O using nonstoichiometric ceria, Science 330 (2010) 1797-1801.

[22] R. R. Bhosale, R. V. Shende, J. Puszynski, Synthesis of nano-sized ferrite materials for H_2 generation from high temperature water-splitting reaction, Proceedings of the 2011 NSTI Nanotechnology Conference and Expo., 1 (2010) 237 – 240.

[23] J.R. Scheffe, R. Jacot, G.R. Patzke, A. Steinfeld, Synthesis, Characterization, and Thermochemical Redox Performance of Hf+4 , Zr+4 , and Sc+3 Doped Ceria for Splitting CO2, The Journal of Physical Chemistry C 117 (2013) 24104-24114.

[24] R. R. Bhosale, R. V. Shende, J. Puszynski, H2 generation from thermochemical water-splitting using sol-gel synthesized ferrites (MxFeyOz, M = Ni, Zn, Sn, Co, Mn, Ce), Proceedings of the 2010 AIChE Annual Meeting (2010).

[25] J.R. Scheffe, D. Weibel, A. Steinfeld, Lanthanum–strontium–manganese perovskites as redox materials for solar thermochemical splitting of H_2O and CO_2, Energy & Fuels 27 (2013) 4250-4257.

[26] R. R. Bhosale, R. V. Shende, M. K. Opoku, J. Puszynski, H2 generation from water-splitting using novel ferrite foam materials, Proceedings of the 2009 AIChE Annual Meeting (2009).

[27] A.H. McDaniel, E.C. Miller, D. Arifin, A. Ambrosini, E.N. Coker, R. O'Hayre, W.C. Chueh, J. Tong, Sr-and Mn-doped $LaAlO_{3-\delta}$ for solar thermochemical H_2 and CO production, Energy & Environmental Science 6 (2013) 2424-2428.

[28] R. R. Bhosale, A. Kumar, A. Ashok, P. Sutar, G. D. Takalkar, M. Khraisheh, F. AlMomani, U. Ghosh, La-Based Perovskites as Oxygen-Exchange Redox Materials for Solar Syngas Production, MRS Advances 2 (2017) 3365-3370.

[29] M. Gálvez, R. Jacot, J. Scheffe, T. Cooper, G. Patzke, A. Steinfeld, Physico-chemical changes in Ca, Sr and Al-doped La–Mn–O perovskites upon thermochemical splitting of CO_2 via redox cycling, Physical Chemistry Chemical Physics 17 (2015) 6629-6634.

[30] R. R. Bhosale, P. Sutar, A. Kumar, F. AlMomani, M. H. Ali, U. Ghosh, S. Al-Muhtaseb, M. Khraisheh, Journal of Renewable and Sustainable Energy 8 (2016) 034702..

[31] F. A. Costa Oliveira, M. A. Barreiros, S. Abanades, A. P. F. Caetano, R. M. Novais, R. C. Pullar, Solar thermochemical CO2 splitting using cork-templated ceria ecoceramics, Journal of CO_2 Utilization 26 (2018) 552-563.

[32] R.R. Bhosale, A. Kumar, F. AlMomani, U. Ghosh, S. Al-Muhtaseb, R. Gupta, I. Alxneit, Assessment of $Ce_xZr_yHf_zO_2$ based oxides as potential solar thermochemical CO_2 splitting materials, Ceramics International 42 (2016) 9354-9362.

[33] M. Hoes, C. Muhich, R. Jacot, G. R. Patcke, A. Steinfeld, Thermodynamics of paired charge-compensating doped ceria with superior redox performance for solar thermochemical splitting of H_2O and CO_2, Journal of Material Chemistry A 5 (2017) 19476-19484.

[34] R.R. Bhosale, A. Kumar, F. AlMomani, I. Alxneit, Sol–gel derived CeO_2–Fe_2O_3 nanoparticles: Synthesis, characterization and solar thermochemical application, Ceramics International 42 (2016) 6728-6737.
[35] C. N. R. Rao, S. Dey, Generation of H_2 and CO by solar thermochemical splitting of H_2O and CO_2 by employing metal oxides, Journal of Solid State Chemistry 242 (2016) 107-115.
[36] R. R. Bhosale. A. Kumar, F. AlMomani, I. Alxneit, Propylene oxide assisted sol–gel synthesis of zinc ferrite nanoparticles for solar fuel production, Ceramics International 42 (2016) 2431-2438.
[37] Y. Hao, C. Yang, S. Haile, Ceria–Zirconia Solid Solutions ($Ce_{1-x}Zr_xO_{2-\delta}$, $x \leq 0.2$) for Solar Thermochemical Water Splitting: A Thermodynamic Study, Chemistry of Materials 26 (2014) 6073-6082.
[38] R. R. Bhosale. A. Kumar, F. AlMomani, U. Ghosh, M. Khraisheh, A comparative thermodynamic analysis of samarium and erbium oxide based solar thermochemical water splitting cycle, International Journal of Hydrogen Energy 42 (2017) 23416-23426.
[39] T. Cooper. J. Scheffe, M. E. Galvex, R. Jacot, G. Patzke, A. Steinfeld, Lanthanum Manganite Perovskites with Ca/Sr A-site and Al B-site Doping as Effective Oxygen Exchange Materials for Solar Thermochemical Fuel Production, Energy Technology 3 (2015) 1130 – 1142.
[40] R. R. Bhosale. Thermodynamic analysis of Ni-ferrite based solar thermochemical H2O splitting cycle for H_2 production, International Journal of Hydrogen Energy (2018) Article In-Press.
[41] C. Muhich, A. Steinfeld, Principles of doping ceria for the solar thermochemical redox splitting of H_2O and CO_2, Journal of Materials Chemistry A 5 (2017) 15578-15590.
[42] R. R. Bhosale, G. D. Takalkar, Nanostructured co-precipitated $Ce_{0.9}Ln_{0.1}O_2$ (Ln = La, Pr, Sm, Nd, Gd, Tb, Dy, or Er) for thermochemical conversion of CO_2, Ceramics International 44 (2018) 16688–16697.
[43] D. Arifin, A. W. Weimer, Kinetics and mechanism of solar-thermochemical H_2 and CO production by oxidation of reduced CeO_2, Solar Energy 160 (2018) 178-185.
[44] R. R. Bhosale, Synthesis and characterization of nanocrystalline $CoFe_2O_4$ – zirconia via propylene oxide aided sol-gel method, Ceramics International 44 (2018) 191-197.
[45] R. R. Bhosale, Thermodynamic efficiency analysis of zinc oxide based solar driven thermochemical H_2O splitting cycle: Effect of partial pressure of O_2, thermal reduction and H_2O splitting temperatures, International Journal of Hydrogen Energy 43 (2018) 14915–14924.
[46] R. R. Bhosale, Solar driven two-step CH_4 reforming and H_2O splitting using Al_2O_3 for Co-production of Al, syngas, and H_2, Solar Energy (2018) Article In-Press.

528408 Effect of Interfacial Heterogeneity on Heavy Oil Desorption

Abstract:

The unconventional oils (e.g., heavy oil, oil/tar sands bitumen) are considered as important alternative to traditional crude oil. Separation of heavy hydrocarbons from mineral surfaces is the key step for unconventional oil production (He, L. et al. Chem. Soc. Rev., **2015**). The interfacial heterogeneities, especially the mineral type and oil composition, play an important role in the desorption process.

In this study, silica and muscovite are applied as two typical host solid surfaces to investigate the effect of mineral type on oil desorption. Results of liberation by visualization measurements demonstrated that it was much slower for oil to liberate from silica surface than that from muscovite surface. This result was further confirmed by surface force measurements. Molecular dynamic (MD) simulation was also conducted to provide detailed profiles of molecular scale structure, dynamics, and energetics of the adsorbed model heavy oil molecules on the mineral surfaces, which was highly consistent with the above experimental results.

Another essential factor affecting the separation efficiency is the oil composition. The petroleum are divided into SARA fractions (saturates, aromatics, resins, asphaltenes) to investigate their desorption behaviors on mineral surfaces (silica, kaolinite and calcium carbonate). The Quartz Crystal Microbalance with Dissipation (QCM-D) tests showed that heavier oil fractions (resins and asphaltenes) exhibited less desorption, caused more significant wettability alteration to the mineral surfaces, and dominated in determining the physicochemical and desorption properties of bitumen. Besides, kaolinite and calcium carbonate showed closer affinity with heavy oil fractions through polar and chemical interactions.

The findings in this study provide fundamental understanding of the effects of mineral type and oil composition on the heavy oil separation, allowing potential insights to the development of enhanced oil recovery.

530099 Synthesis and Application of Amino Acid Ionic Liquid-Based Deep Eutectic Solvents for Oil-Carbonate Mineral Separation

Abstract:

Synthesis and application of amino acid ionic liquid-based deep eutectic solvents for oil-carbonate mineral separation

Ning KANG[a], Lin HE[a,b], Hong SUI[a,b,c], Zisheng ZHANG[a,d]

[a] School of Chemical Engineering and Technology, Tianjin University, Tianjin 300072, China.

[b] Collaborative Innovation Center of Chemical Science and Engineering(Tianjin), 300072, China.

[c] National Engineering Research Centre of Distillation Technology, Tianjin, 300072, China.

[d] Department of Chemical and Biomedical Engineering, University of Ottawa, Ottawa, ON K1N 6N5, Canada.

The Indonesian asphalt rock is one of the carbonate unconventional oil ores, which has been attracting increasing attention from the industry due to its huge geological reserves (~1 billion barrels in situ)(Li et al., 2012) . To unlock this kind of petroleum ore, different methods have been proposed, and the solvent extraction (Nikakhtari et al., 2013, He et al., 2015) has been considered to be a potential method for the ore processing due to the following advantages: high efficiency, low or no water consumption, mild operation conditions and good applicability for oil-wetted ores. Recently, much of the attention has been paid on developing new solvents or innovative process aids with high extraction efficiency. The ionic liquids with unique properties have been applied to enhance the solvent extraction process. It is found that the application of ionic liquids together with solvents not only increases the oil recovery, but also improves the product quality with negligible fines being entrained. (Painter et al., 2009, Painter et al., 2010). However, the high cost and complicated synthesis process of the present involved ILs restrict their further applications in oil sands extraction.

In the present study, to allow the process much more cost-saving, cheaper and environment friendly chemical agents, deep eutectic solvents (which are considered as potential alternatives to traditional ionic liquids) has been synthesized by protonated Tryptophan salt and urea (i.e. $TrpBF_4/U$). When mixing with the ethanol and toluene, the $TrpBF_4/U$ would form a non-aqueous and surfactant-free micro-emulsion with the emulsion droplets less than 7 nm. Owing to the low interfacial tension (0.95mN/m), this newly prepared $TrpBF_4/U$ micro-emulsion was found to perform well in separating the extra-heavy oil from carbonate asphalt rocks (up to 93% of bitumen recovery in a single extraction), resulting in better oil quality with less solid entrained (1.07 mg of solids/mL of solvent) compared with the previous ionic liquids (i.e., $[Emim]BF_4$). Results of water drop penetration time and elemental composition of solid surfaces also showed $TrpBF_4/U$ plays an important role in interfacial alternation at the oil and mineral surfaces. It is also found that the $TrpBF_4/U$ micro-emulsion could be recycled at least 4 times without sacrificing the extraction efficiency.

These findings of the $TrpBF_4/U$ micro-emulsion showed promising application in unconventional oil separation. It is also believed this kind of materials would be applied in other related extraction, such oil-contaminated soil remediation, bio-fuels extraction, deinking, etc.

References:

He, L., F. Lin, X. G. Li, H. Sui and Z. H. Xu, 2015. Interfacial sciences in unconventional petroleum production: from fundamentals to applications. Chemical Society Reviews 44, 5446-5494.

Li, X., L. He, G. Wu, W. Sun, H. Li and H. Sui, 2012. Operational Parameters, Evaluation Methods, And Fundamental Mechanisms: Aspects of Nonaqueous Extraction of Bitumen from Oil Sands. Energy & Fuels 26, 3553-3563.

Nikakhtari, H., L. Vagi, P. Choi, Q. Liu and M. R. Gray, 2013. Solvent screening for non-aqueous extraction of Alberta oil sands. Canadian Journal of Chemical Engineering 91, 1153-1160.

Painter, P., P. Williams and A. Lupinsky, 2010. Recovery of Bitumen from Utah Tar Sands Using Ionic Liquids. Energy & Fuels 24, 5081-5088.

Painter, P., P. Williams and E. Mannebach, 2009. Recovery of Bitumen from Oil or Tar Sands Using Ionic Liquids. Energy & Fuels 24, 1094-1098.

527295 Geothermal Sludge-Derived Calcined Sodium Silicate As Heterogeneous Catalyst for Biodiesel Production from Waste Cooking OIL

Abstract:

GEOTHERMAL SLUDGE-DERIVED CALCINED SODIUM SILICATE AS HETEROGENEOUS CATALYST FOR BIODIESEL PRODUCTION FROM WASTE COOKING OIL

Ichsan Dwi Nugraha

Department of Chemical Engineering, Institut Teknologi Sepuluh Nopember

ichsandwinugraha@gmail.com

Abstract - The depletion of world petroleum reserves and increased environmental concerns such as polluting air by emissions of sulfur dioxides, carbon dioxides, particulate matter, and other gases has resulted to increased research in alternate fuels and renewable source of energy. Trans-esterification process used to produce biodiesel from waste cooking oil, bears a large potential for the substitution of fossil diesel. Biodiesel is proved to be the best replacement for diesel because of its unique properties like significant reduction in greenhouse gas emissions, non-sulfur emissions, non-particulate matter pollutants, low toxicity and biodegradability. Therefore, the employment of waste oils as feedstock for biodiesel production are presented as the best alternative. The use of solid base catalyst in biodiesel production is becoming preferable because of their superiority advantages over the conventional homogeneous catalyst. Calcined sodium silicate is applied to catalyze trans-esterification reaction of vegetable oil with methanol. A strong base catalyst of sodium silicate synthesized from geothermal sludge of Dieng Geothermal Power Plant (Indonesia), which has 55% of silica content by gelation method. The factors affecting the process parameters reported are studied and the point of interest focuses on their catalyst both qualitative and quantitative scope, methanol to oil ratio, and reaction temperature. Sodium silicate catalyst was activated by calcination process at 400°C, temperature ramp of 20°C/min, and 3 hours detention time. The trans-esterification was carried out at various temperature in the range of 50°C - 70°C with a methanol-oil molar ratio of 9:1 and 3% (w/w) of catalyst for 60 minutesâ€™ reaction time. To investigate kinetics of reaction, the sample were taken consecutively during the reaction after pouring into the reactor. Experimental results showed that the sodium silicate was very active in the trans-esterification of vegetable oil with methanol. Reaction temperature at 60Â°C was sufficient to reach a conversion level as high as above 94% in a relatively short reaction period. Results of the present work suggested that sodium silicate synthesized from local resources of geothermal sludge would become potential solid base catalyst in biodiesel synthesis.

PERFORMANCE OF MICROEMULSIONS AS AN ALTERNATIVE FUEL IN CONSTANT SPEED DIESEL ENGINE

Iyman Abrar and Ashok N. Bhaskarwar
Indian Institute of Technology Delhi, New Delhi, India

Abstract

There is an urgent need for sustainable alternative fuels due to the limited stocks of and increasing demand for fossil fuels, as well as evermore increasingly-stringent environmental regulations. Microemulsions are thermodynamically stable alternatives which could be directly used in the engine without any modification. Our current work attempts to examine the performance of different microemulsion fuels in a diesel engine. In the current investigation, an alcohol was used which acted both as a surfactant as well as a co-surfactant, thereby making the entire process facile and economical. Alcohols are oxygenates, and reduce the CO and HC emissions. The higher latent heat of alcohol reduces the temperature inside the combustion chamber, which, in turn, reduces the NO_x- and smoke-emissions as well. Various microemulsion compositions with constant diesel percentages (60% and 70%) were formulated, and characterized as per ASTM D975. The microemulsions formulated with 70% diesel satisfied the ASTM standards, while some of the compositions with 60% diesel showed reduced cetane index. The percentage change in the property of fuel 'ε' was calculated as $[(\varepsilon_{diesel}-\varepsilon_{microemulsion})/\varepsilon_{diesel}]\times100$. Preliminary engine studies revealed that the microemulsion fuel containing 70% diesel had 15.55% higher brake thermal efficiency, and 5.04% lower brake specific fuel consumption compared to that of neat diesel, even at 8.31% reduction in calorific value. Hence, the microemulsions can be used as an environment-friendly alternative to diesel.

Keywords

Alternative fuel, alcohol, microemulsion, ASTM, diesel engine

Introduction

Worldwide fossil fuels are the primary source of energy. Their limited production and increased consumption are resulting in a substantial increase in costs. Apart from that, the fossil fuels are a major contributors to global warming and climatic changes. Among the fuels for mass transportation, diesel is used in much higher amounts than petrol due to their better power conversion efficiency and better fuel economy in compression ignition engines. They emit a lot of pollutants, mainly particulate matters (PM) which are carcinogenic, and NO_x which are responsible for acid rain, smog and ozone formation at the ground level (1). Diesel also contributes to the emissions of carbon dioxide (CO_2), carbon monoxide (CO), black smoke, and sulphur oxides (SO_x). Standard specifications of diesel fuel oil (Nos. 1 and 2) are given by ASTM (2).

Biodiesel is considered to be the most sustainable alternative to diesel. It can be obtained from the transesterification of vegetable oil and animal fat. It is biodegradable, non-toxic, and oxygenating fuel. The presence of oxygen in the fuel itself allows for a more complete combustion, thus, reducing CO and HC emissions (3). However, the calorific value of biodiesel is lower than that of diesel, causing an increased fuel consumption and a reduced thermal efficiency (4). The presence of nitrogen in the vegetable oil causes higher NO_x emissions from biodiesel (5). It is also more viscous than diesel, and has a lower oxidation stability (6). It also clogs the fuel filter and fuel lines at low temperatures, causing an incomplete combustion and problems in starting the engine (7).

Microemulsion is another promising alternative to diesel fuel, which is preferred over other alternatives as no reaction is involved. The term *microemulsion* was coined by Schulman for an optically isotropic dispersion of oil and water (8). They are thermodynamically stable colloidal dispersions, with the dispersed particle size smaller than one-fourth of the wavelength of visible light (9). The microemulsion fuels generally consist of water-in-oil reversed micellar structure, or sometimes the bicontinuous lamellar structure. They are sensitive to temperature since the curvature of the film changes at a critical point known as phase-inversion temperature (PIT) (10). Water droplets in microemulsions are surrounded by a film of diesel fuel, and evaporate rapidly at high temperatures in a combustion chamber, causing microexplosion or secondary atomization (11). Water also causes a reduction in NO_x emissions, due to its high latent heat which reduces the temperature inside the combustion chamber (12). The presence of oxygenates, such as alcohols, reduce the PM by 15-20% for an oxygen content of 2-3% (13).

Increase in density and viscosity with increase in percentage of water in microemulsions has been reported for diesel-based microemulsions stabilized by ethoxylated nonylphenol (12). However, the fuel consumption improved, along with reduction in NO_x and smoke emissions. In another study, ethanol-diesel microemulsions stabilized by biodiesel and span-80 showed reduction in fuel consumption, and increase in thermal efficiency due to the presence of oxygenates, which promote combustion (14). Effect of different types of surfactant on the microemulsion region had also been studied, and it was observed that a hydrophilic non-ionic surfactant expanded the microemulsion region, while a hydrophobic non-ionic surfactant narrowed it (15). More recently, it has been observed that an alcohol can act both as a surfactant and a co-surfactant, thus eliminating the need for an additional surfactant for the formulation of alcohol-diesel-water microemulsions (16). These formulated microemulsions have properties as per ASTM standards, and hence could possibly be used as a diesel replacement. The current work investigates the effect of different alcohol percentages on the properties of such surfactant-free microemulsion fuels at constant diesel percentages at two levels of 60 % (v/v) and 70 % (v/v).

Experimental work

Materials

Chemicals (Analytical Reagent (AR) grade), used for the formulation of microemulsion fuels and their testing as per ASTM, were purchased from Fisher Scientific, while deionized water was procured from Organo Biotech Laboratory. High-speed-diesel (HSD), used as petrodiesel, was obtained from the sales end of Indian Oil Corporation Limited (IOCL).

Methods

Known quantities of alcohol and diesel were first mixed, and then water was added. The mixture was stirred for 2-3 min using a magnetic stirrer. A red laser beam was then passed through the microemulsions. The microemulsion was confirmed if the beam passed through it without getting scattered, while emulsion was present if the beam got scattered with a much-reduced intensity (17). Eight different compositions were chosen from the microemulsion domains, at constant diesel percentages of 60 and 70 % (v/v), with a small percentage of water (0.5–2 % (v/v)), and 29–39.5 % (v/v) alcohol for testing as per ASTM D975.

The calorific values of the fuels formulated were determined using an isoperibol calorimeter (Make Leco, Model AC600). Specific gravities were measured with the help of a density bottle (Make Borosil). The viscosities were measured with a viscometer (Make Hindustan Apparatus, Model HAMCO 50 A-50) kept in a constant-temperature bath (Make Thermotech, Model TC4000N) at 40°C. The cetane index was calculated from API gravity (Make Alla France, Model 3503LC085/15/B) and mid-boiling point temperature (Make Welkins, Model Welkins D.A.). The cloud points were measured with a cloud-point apparatus (Make Welkins, low temperature with a range of -38°C to +50°C). Ash content was estimated using a high-temperature furnace (Make Selec, Model NSW103), while the sulfur content was determined using wavelength dispersive X-Ray Fluorescence (WDXRF) spectrometer (Make Rigaku, Model Super Mini). Ramsbottom carbon residue was checked by Ramsbottom carbon residue apparatus (Make Khera, Model Khera RCR), while water and sediments were measured using an oil centrifuge (Make Eltek, Model OC 2F), the copper strip corrosion rating by copper strip corrosion test apparatus (Make Scientech Instruments, Model SE-297), and the distillation temperature by a batch-distillation unit (Make Welkins, Model Welkins D.A).

One of the microemulsion samples was also run in a CI engine (Make Kirloskar, Model TAF-1), attached to an alternator (Make Wilson) and a loading panel to check the fuel consumption and efficiency as compared to those of neat diesel.

Results and discussion

Calorific value

Calorific value is the amount of heat released per unit mass of the fuel while undergoing complete combustion. The calorific values of the fuel samples were measured in an isoperibolic bomb calorimeter, as per ASTM D240. There is an increase in the calorific value of a microemulsion fuel as compared to the weighted mass basis in both 60% and 70% diesel concentrations, as shown in figure 1. Moreover, the increase is more in case of 60% diesel, indicating that the microemulsions have modified reaction pathways due to the presence of alcohol and water, leading to more complete combustion, and thereby increasing the calorific value of the fuel.

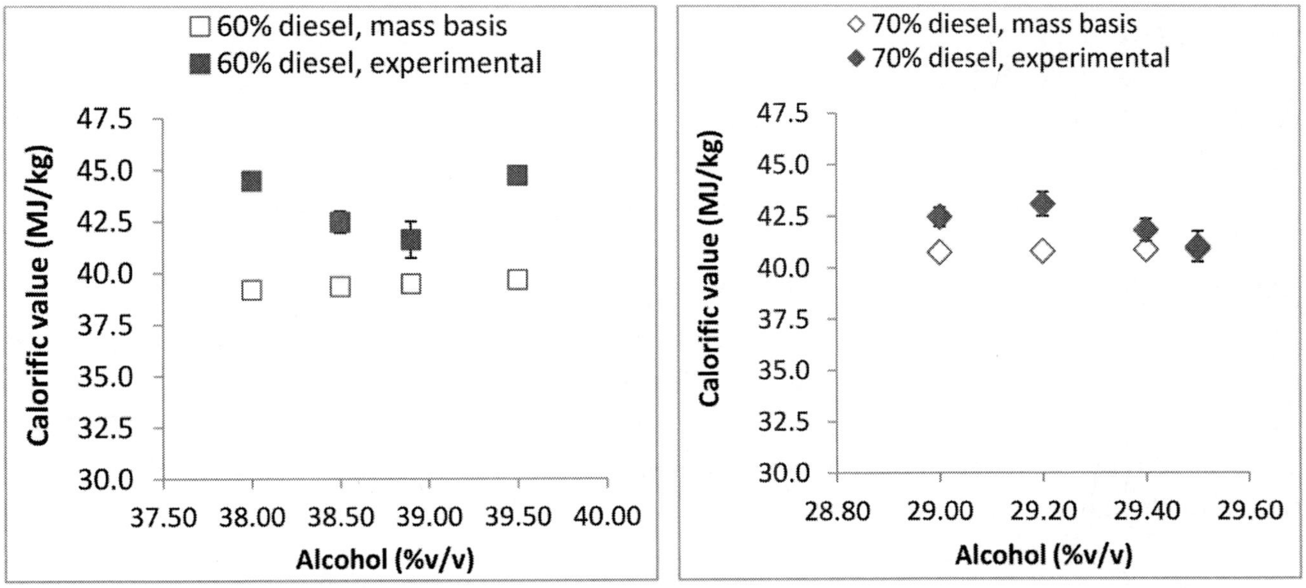

Figure 1. Variation in the calorific values of microemulsion fuels, at 60 % (v/v) and 70 % (v/v) diesel, with the change in percentage of alcohol.

Viscosity

Viscosity is an important parameter which determines proper atomization and combustion of the fuel. The viscosities of the microemulsions were measured as per ASTM D445. The viscosities of microemulsions were lesser as compared to the weighted averages, as shown in figure 2, possibly due to the presence of alcohol, which could cause slipping of fluid layers over each other, resulting in the reduced viscosities. The viscosities of all the microemulsions were observed to be within ASTM-specifications of 1.9 — 4.1 mm^2/s.

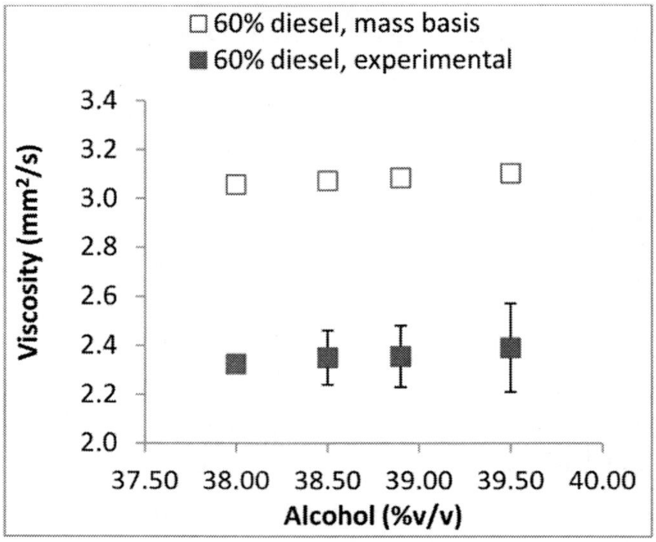

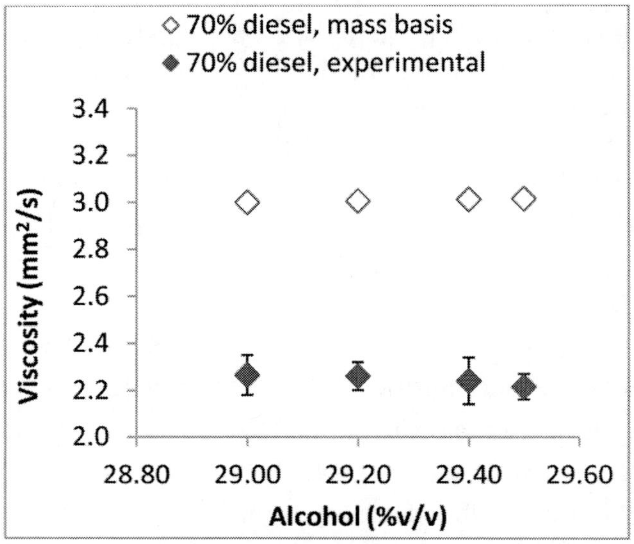

Figure 2. Variation in the viscosity of microemulsion fuels, at 60 % (v/v) and 70 % (v/v) diesel, with the change in percentage of alcohol.

Cetane index

The calculated cetane index gives an estimate of cetane number of the fuel. It was calculated as per ASTM D976 using API gravity and mid-boiling point temperature of the microemulsion fuel, using equation (1).

$$Calculated\ cetane\ index = -420.34 + 0.016G^2 + 0.192G \log M + 65.01(\log M)^2 - 0.0001809M^2 \quad ...(1)$$

, where G is API gravity and M is mid-boiling temperature, °F.

No clear trend was observed in the cetane index of microemulsions versus alcohol percentage, as shown in figure 3. However, the microemulsions with 60% diesel possibly cannot be used in the engine due to the low cetane index. Furthermore, the cetane index for microemulsions with 70% diesel is higher than the weighted averages, indicating a lower ignition delay, and more complete combustion caused by the modified reaction pathways.

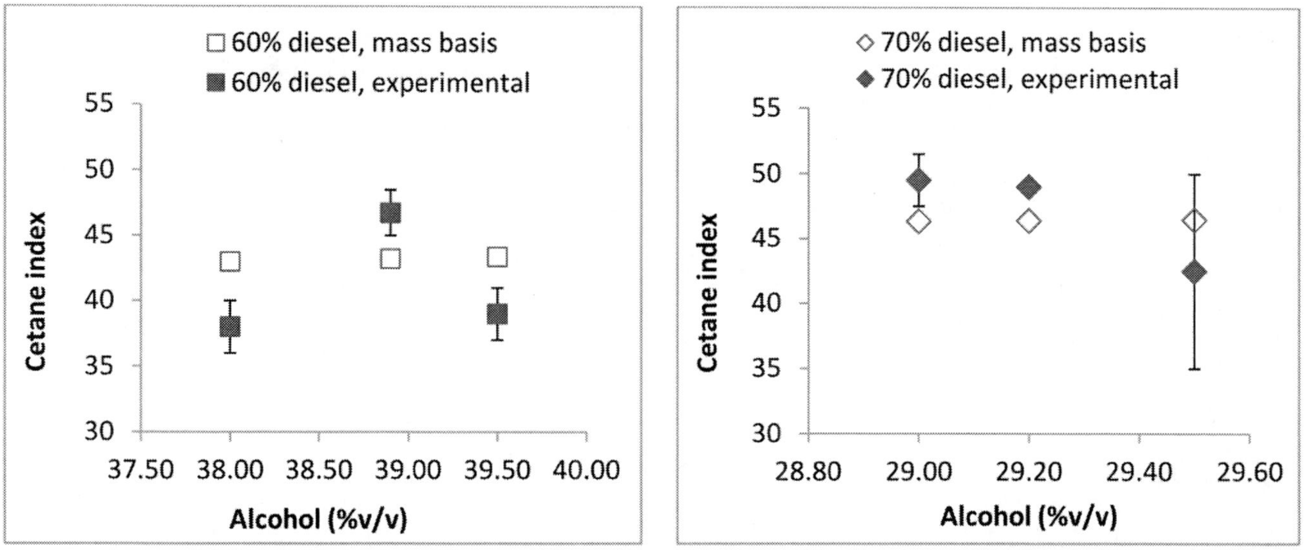

Figure 3. Variation in the cetane index of microemulsion fuels, at 60 % (v/v) and 70 % (v/v) diesel, with the change in percentage of alcohol.

Cloud point

The cloud points were measured using the cloud-point apparatus, as per ASTM D2500. ASTM D975 does not specify the lowest limits of the temperature for cloud point, while IS 1460 specifies cloud points to be 3°C for winter and 15°C for summer fuels. The weighted averages have not been calculated as the microemulsions were expected to start to cloud with the fuel component having the highest cloud point, i.e. water in this case. The cloud points of all the microemulsions, however, were lower than 0°C, as shown in figure 4, and close to that of diesel for 70%-diesel microemulsion compositions. Moreover, no phase separation was observed in the microemulsions up to -10°C, indicating stable microemulsion formation.

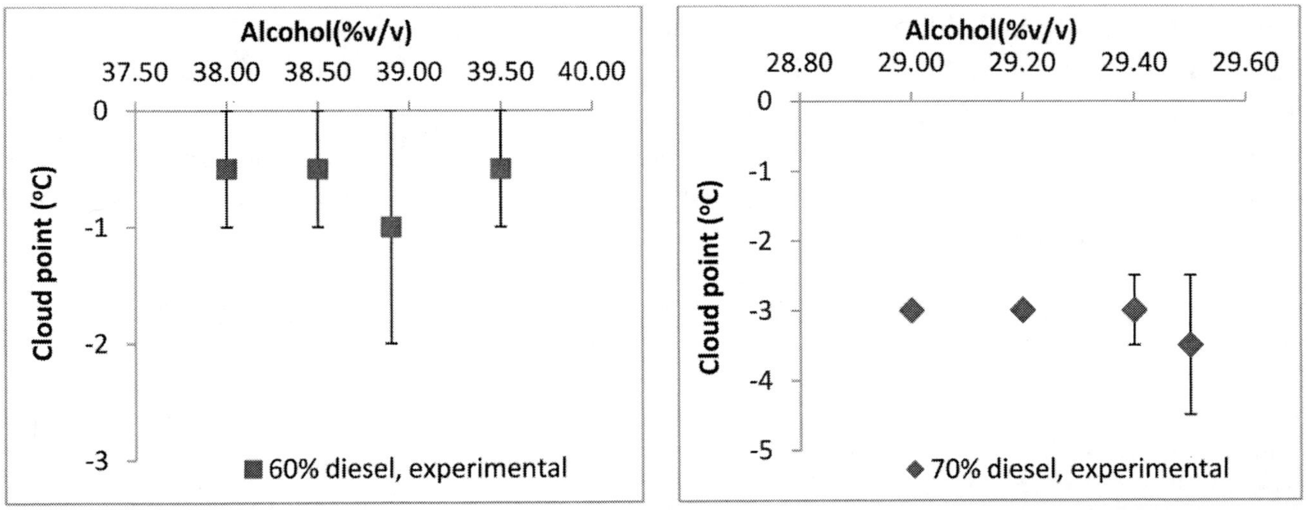

Figure 4. Variation in the cloud point of microemulsion fuels, at 60 % (v/v) and 70 % (v/v) diesel, with the change in percentage of alcohol.

Ash content

Ash content was estimated using a high-temperature furnace at 700-800°C, as per ASTM 482. The ash contents of all the microemulsions were within the permissible limits of 0.01% mass, as shown in figure 5. The ash content was smaller for the microemulsions with 60 % (v/v) diesel, perhaps due to the more complete combustion caused by oxygenating nature of alcohol.

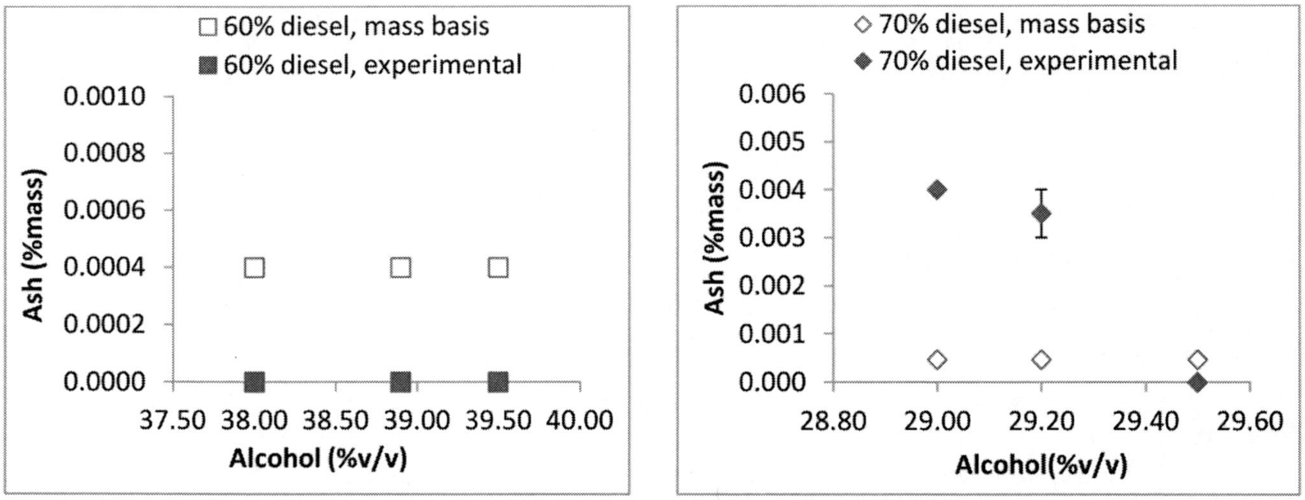

Figure 5. Variation in the ash percentage of microemulsion fuels, at 60 % (v/v) and 70 % (v/v) diesel, with the change in percentage of alcohol.

Sulfur content

The maximum allowable sulfur content is 15 ppm for S15, 500 ppm for S500, and 5000 ppm for S5000 as per ASTM D975, and 50 ppm as per IS 1460. Since the diesel stock was procured from IOCL outlet, which specifies the sulfur content to be less than 50 ppm, IS 1460 has been followed for sulfur measurements. Sulfur content was measured using a wavelength dispersive X-ray fluorescence spectrometer, as per ASTM D2622, which has an oxygen tolerance of 2.8%. Since the microemulsions have a higher oxygen content due to the presence of water and alcohol added, this oxygen interferes with the measurements, resulting in a higher estimated sulfur content as compared to the weighted average, as shown in figure 6. Even after the interference, all of the formulated microemulsion samples showed the sulfur content of less than 50 ppm. Moreover, the microemulsions with 60% diesel had a lower level sulfur content as compared to that with 70% diesel.

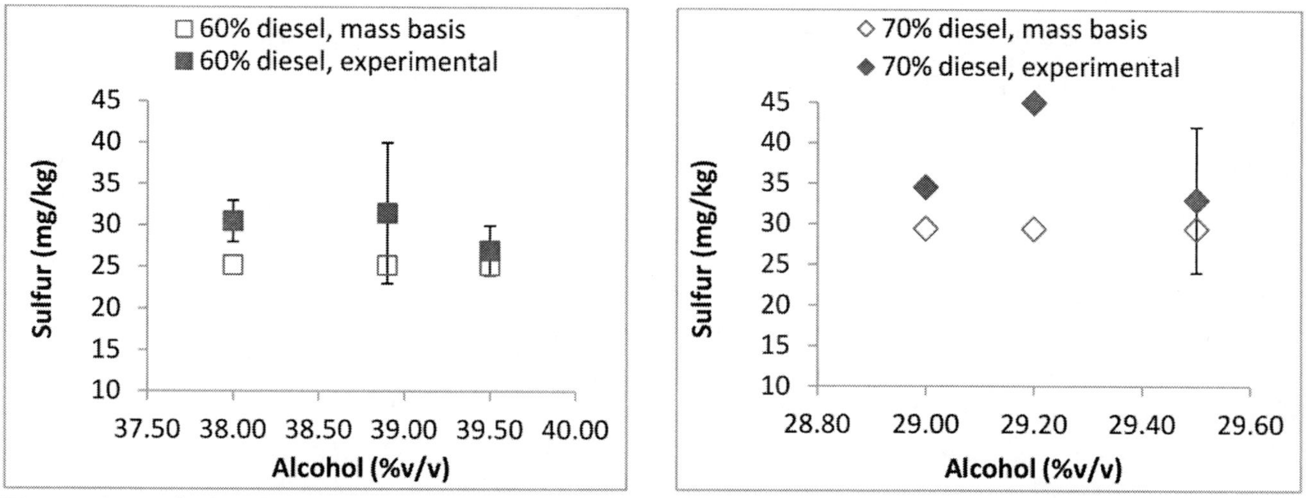

Figure 6. Variation in the sulfur content of microemulsion fuels, at 60 % (v/v) and 70 % (v/v) diesel, with the change in percentage of alcohol.

Ramsbottom carbon residue

Ramsbottom carbon-residue apparatus was used to measure the carbon residue on 10% distillate at 550°C, as per ASTM D524. The carbon residue should be lower than 0.35 % mass as per ASTM D975, as it can form deposits in the combustion chamber. All the samples had the residue within the permissible limits, as shown in figure 7.

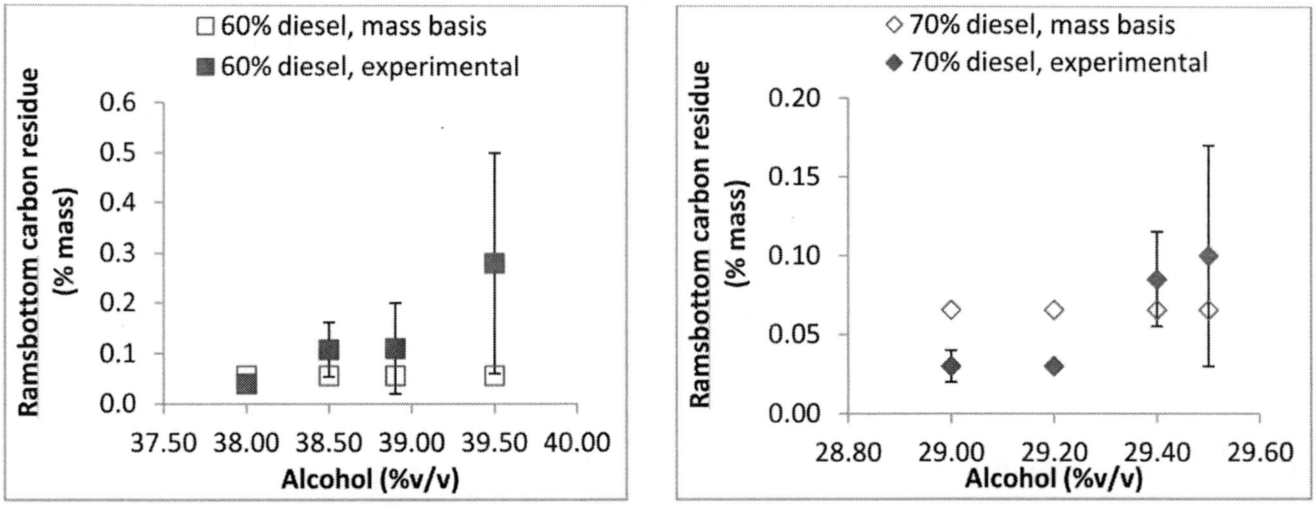

Figure 7. Variation in the percentage of Ramsbottom carbon residue of microemulsion fuels, at 60 % (v/v) and 70 % (v/v) diesel, with the change in percentage of alcohol.

Water and sediments

The water and sediments percentage of the fuel samples were measured as per ASTM D2709 at a relative centrifugal force of 800. All the microemulsions had water and sediments percentage lower than weighted averages which indicate that the water is tightly bound, and cannot be separated even by centrifugation.

Copper strip corrosion test

The copper-strip corrosion rating indicates the tendency of the fuel to corrode copper, brass, or bronze parts of the engine. Copper-strip corrosion rating was measured at 50°C for three hours as per ASTM D130. All the samples showed a rating of No.1 which indicated that microemulsions were not corrosive to the engine.

Fuel consumption and thermal efficiency

Preliminary engine testing was carried out for microemulsion sample 4, with 70% diesel and 29.5% alcohol, and compared against neat diesel. The percentage change in the property of fuel, 'ε', was calculated using the following equation.

$$[|(_{diesel} - _{microemulsion})|/_{diesel}] \times 100 \qquad \ldots(2)$$

The fuel consumption decreased, and the thermal efficiency increased with an increase in the load, as the temperature and turbulence inside the combustion chamber increased. It was observed that even though the microemulsion sample 4 had a reduction in the calorific value by 8.31%, the fuel consumption was lower by 5.04% while the brake thermal efficiency increased by 15.55% compared to neat diesel, which indicated that the microemulsions caused a more complete combustion inside the combustion chamber.

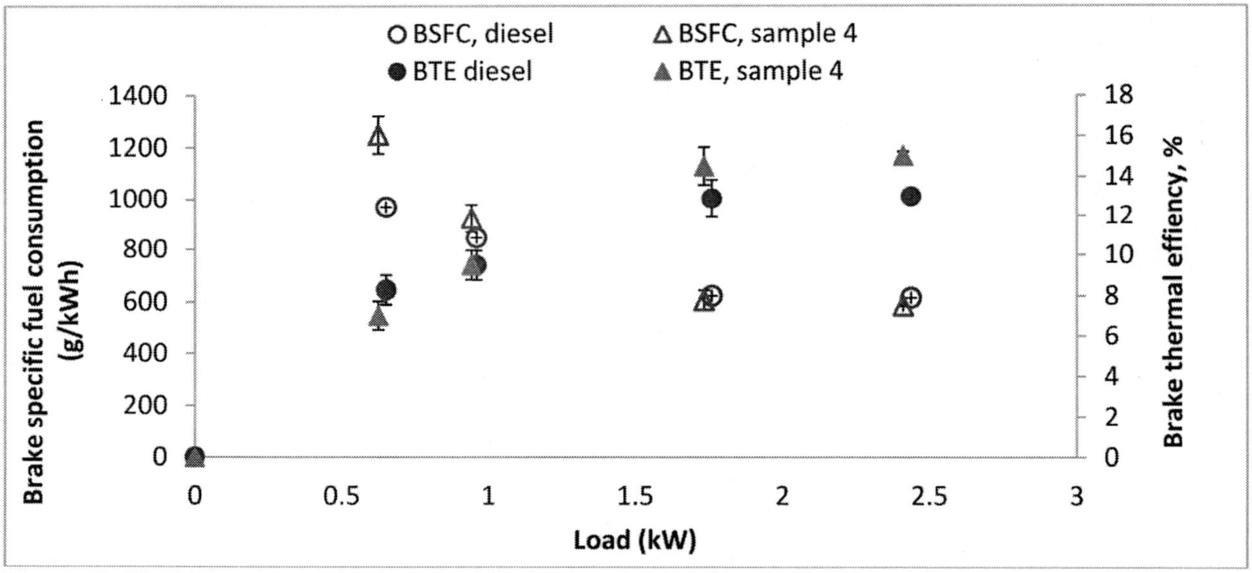

Figure 8. Variation in the brake specific fuel consumption (BSFC) and brake thermal efficiency (BTE) of microemulsion fuel sample 4 (containing 70 % v/v diesel) and diesel with the load.

Conclusion

Eight different microemulsion compositions were tested as per ASTM D975 to determine the effect of alcohol on the properties of microemulsions. It was observed that the microemulsions had higher calorific, and cetane index values as compared to the weighted averages; and lower viscosities, cloud points, ash contents, and percentages of water and sediments. The microemulsions with 70 % (v/v) diesel satisfy all the properties requirement of a diesel fuel as per ASTM D975, while those with 60 % (v/v) diesel had slightly lower cetane indices. Thus, some cetane additives would be needed for the microemulsions with 60% diesel. Furthermore, preliminary engine testing indicated that the microemulsions had a higher thermal efficiency and a lower fuel consumption than with neat diesel. The microemulsions could, thus, be used as a sustainable alternative fuel to conserve the dwindling reserves of petroleum and diesel.

Acknowledgements

The authors would like to thank DST for the DST-INSPIRE (IF150907) fellowship to Ms. Iyman Abrar, and SERB-INDIA for the research grant (EMR/2016/004152) for carrying out the current research work.

References

1. Abrar, I., and Bhaskarwar, A. N. (2018) Formulation and extension of diesel-based microemulsion fuels for compression ignition engines. *INAE Lett.* 3 (1): 33–39.
2. ASTM: D975-15b. (2015) *Standard specification for diesel fuel oils*. Available at: http://www.astm.org/Standards/D975.htm.
3. Panneerselvam, N., Murugesan, A., Vijayakumar, C., Kumaravel, A., Subramaniam, D., and Avinash, A. (2015) Effects of injection timing on bio-diesel fuelled engine characteristics - An overview. *Renew. Sustain. Energy Rev.* 50: 17–31.
4. Aldhaidhawi, M., Chiriac, R., and Badescu, V. (2017) Ignition delay, combustion and emission characteristics of diesel engine fueled with rapeseed biodiesel – A literature review. *Renew. Sustain. Energy Rev.* 73: 178–186.
5. Ramkumar, S., and Kirubakaran, V. (2016) Review on admission of preheated vegetable oil in C.I. engine. *Indian J. Sci. Technol.* 9 (2): 1–11.
6. Saluja, R. K., Kumar, V., and Sham, R. (2016) Stability of biodiesel – A review. *Renew. Sustain. Energy Rev.* 62: 166–181.
7. Monirul, I. M., Masjuki, H. H., Kalam, M. A., Zulkifli, N. W. M., Rashedul, H. K., Rashed, M. M., Imdadul, H. K., and Mosarof, M. H. (2015) A comprehensive review on biodiesel cold flow properties and oxidation stability along with their improvement processes. *RSC Adv.* 5 (105): 86631–86655.
8. Schulman, J. H., Stoeckenius, W., and Prince, L. M. (1959) Mechanism of formation and structure of micro emulsions by electron microscopy. *J. Phys. Chem.* 63 (10): 1677–1680.
9. Goering, C. E., Schwab, A. W., Campion, R. M., and Pryde, E. H. (1983) Soyoil-ethanol microemulsions as diesel fuel. *Am. Soc. Agric. Eng.* 26 (6): 1602–1605.
10. Shinoda, K., and Saito, H. (1969) The stability of O/W type emulsions as functions of temperature and the HLB of emulsifiers: The emulsification by PIT-method. *J. Colloid Interface Sci.* 30 (2): 258–263.
11. Acharya, B., Guru, P. S., and Dash, S. (2014) Tween-80–n-butanol–diesel–water microemulsion system—A class of alternative diesel fuel. *J. Dispers. Sci. Technol.* 35 (10): 1492–1501.
12. Neto, A. A. D., Fernandes, M. R., Barros Neto, E. L., Dantas, C. T. N., and Moura, M. C. P. A. (2011) Alternative fuels composed by blends of nonionic surfactant with diesel and water: Engine performance and emissions. *Brazilian J. Chem. Eng.* 28: 521–531.
13. Chandra, R., Mishra, T. N., and Bhattacharya, T. K. (2006) Exhaust emission characteristics of alcohol-diesel microemulsions in a low bhp constant speed compression ignition engine. *J. Inst. Eng. Agric. Eng. Div.* 87: 15–19.

14. Mehta, R. N., Barad, J. M., Chakraborty, M., and Parikh, P. A. (2012) A stability and performance study of ethanol-diesel microemulsion fuel. *Pet. Sci. Technol.* 30: 159–169.
15. Abrar, I., and Bhaskarwar, A. N. (2017) Microemulsion fuel for compression ignition engines: formulation and extension of microemulsion domains. *6th International Conference on Advances in Energy Research*, Mumbai, India.
16. Abrar, I., and Bhaskarwar, A. N. (2016) A microemulsion fuel composition and method of formulating. 201611039919.
17. Attaphong, C., and Sabatini, D. A. (2013) Phase behaviors of vegetable oil-based microemulsion fuels: The effects of temperatures, surfactants, oils, and water in ethanol. *Energy & Fuels* 27 (11): 6773–6780.

518407 Experimental Study of a Fluidized Bed Reactor for Obtaining Biodiesel from *Jatropha curcas* Oil By Means of Immobilized Enzymes: Stage 1: Characterization of Lipase

Abstract:

In this research project we propose to study the main hydrodynamic variables of a bioreactor that intervene in the degree of conversion in the transesterification reaction of vegetable oil from Jathropha Curcas L. to produce biodiesel using an immobilized enzyme (lipase) as a biocatalyst. The hydrodynamic parameters to be studied will be: the minimum fluidization velocity, velocity of expansion and porosity of the bed; and the operating variables will be: temperature, oil feed flow, type of biocatalyst, biocatalyst concentration. In addition, the type of immobilization, the physical characteristics of the support and the degree of conversion will be determined. The conversion degree data obtained will be compared against experimental data in a continuously stirred batch reactor.
It will work with vegetable oil of J. Curcas L provided and characterized by the CEPROBI - IPN. In this first stage the results of the characterization of the lipase obtained from the protein fraction of the Jathropha (Circular Dichroism Analysis, AFM, MET, SEM, Confocal, SDS PAGE, peptide trace and RMN H and C) and its comparison with a commercial lipase will be shown. Advances in the tests of immobilization via adsorption and encapsulation in commercial supports will be shown. The one that provides the best results of immobilization of the enzyme will be chosen as support.
The degrees of conversion were determined qualitatively by the formation of ethyl esters (biodiesel) at the end of the process, using gas chromatography coupled to masses.
The planning of experiments will be done by a 2n statistical factorial model with two levels and "n" factors. The response variable will be the degree of conversion.

537664 Alcohols/Mixed Oxygenates Conversion to Higher Ketones over Multifunctional Mixed Oxide Catalysts

Abstract:

There is a need to replace fossil derived chemicals and fuel by using renewable materials and processes. Ethanol and Acetone, Butanol, Ethanol (ABE) mixture can be efficiently produced from biomass fermentation. Ethanol/ABE could be converted to higher ketones in thermochemical pathway using heterogeneous catalysts in a single step process. Transition metals impregnated on carbon supports had shown to promote one-step synthesis of ketones. Catalysts with more stable supports could be synthesized to promote higher conversion and selectivity towards ketones. In our experiment mixed oxides of manganese, zinc, zirconia was used as support for ketones synthesis. The catalyst was tuned with addition of transition metals to promote specific pathway with higher conversion of Ethanol/ABE with 70%-80% selectivity towards C_{5+} ketones. Acidic and basic properties of the catalyst promotes dehydrogenation, aldol condensation, dehydration and hydrogenation reactions. The condensation of acetone with acetaldehyde to form pentanone rather than the self-condensation improves the final product yield and reduces the overall carbon loss. This presentation will discuss the robustness of tuned catalysts under various conditions, product yields and the reaction mechanisms.

533813 Single-Step Co-Synthesis of Methanol, Dimethyl Ether and Dimethyl Carbonate from Biomass Derived Syngas

Abstract:

Gasification of biomass to syngas (CO, CO_2 and H_2) is attractive biomass utilization route due to high throughput, fast reaction rates, and low separation costs. Moreover, the *syngas* from biomass paves way for the greener synthesis route for important industrial intermediates, oxygenates and solvents such as methanol, dimethyl ether (DME) and dimethyl carbonate (DMC). The production of DME and DMC are industrially mature technologies that are carried out as a two-step process via methanol dehydration and methanol oxy-carbonylation respectively. However, the single-step co-synthesis of these chemicals from syngas has not been reported. Moreover, the enhanced utilization of CO_2 requires the use of novel catalytic systems for improved yields and selectivity.

In this research, mixed-oxide supported Cu and In_2O_3 catalysts have been developed and specifically the support-active component effects have been investigated. CeO_2-ZrO_2 mixed oxides have been used as the supports in this study mainly because of their varying acidity and basicity that influence the product distribution. The chemical synthesis experiments using the developed catalysts were carried out in the range of 100-300°C, 20-50bar and a space velocity of 23000-60000 hr^{-1} in a tailor-made high pressure packed bed reactor with an online GC-TCD/FID/FPD analyser. In-depth catalyst characterization was conducted including the in-situ synchrotron powder diffraction, X-ray absorption spectroscopy, in-situ diffuse reflectance infrared Fourier transform (DRIFTS) and scanning transmission electron microscopy (STEM).

Several optimization experiments were carried out including the influence of process parameters such as temperature, pressure, space velocity and CO co-feeding in addition to catalyst parameters such as support composition (Ce/Ce+Zr=0,0.2,0.5,0.8,1) and active metal loading. The products from the synthesis include DME, methanol, DMC and methane. Support composition has a strong influence on the yields and product selectivity. The novel catalysts developed in this study exhibit exceptional performance in terms of yields and stability compared to the ones reported in literature. CeO_2 support was found to be the most efficient for both the Cu ($0.47 g_{MeOH}\ h^{-1}\ g_{Cu}^{-1}$) and In_2O_3 ($1.5\ g_{MeOH}\ h^{-1}\ g_{In}^{-1}$) catalysts. The results from this study contribute to significant process improvement through the design of a versatile catalyst enabling co-production of methanol, DME and DMC.

537553 Insight into the Reactions of Upgrading Bio Aldehydes to High Value Aromatic Precursors

Abstract:

Terephthalic acid and phthalic anhydride are high value and high market commodity chemicals, used in polyester and plasticizers industry. These chemicals are currently produced from oxidation of p-xylene and o-xylene that are derived from the crude oil based resources. In this work, 4-methylbenzaldehyde and 2-methylbenzaldehyde, precursor of terephthalic acid and phthalic anhydride, were synthesized from biomass derived acetaldehyde through catalytic aldol condensation. The conversion of acetaldehyde to methyl benzaldehyde reaction takes place in two step reaction mechanism. The first step is the acetaldehyde condensation to form crotonaldehyde, and the second step is the conversion of crotonaldehyde to methylbenzaldehyde. Crotonaldehyde is an Î±,Î²-unsaturated C4 aldehyde, which contains two positions (C=C and C=O) to form the C-C bond. Therefore, the position of the C-C bond formation plays major factor in determining the product selectivity. In this research, various acid-base catalyst with different strengths, such as hydrotalcite and ZnZr, were investigated to understand how the nature of the basic and acid site control the reaction pathway to improve the product selectivity. In addition, several reactants with different unsaturated bond were incorporated as model reactants for reaction pathway study. The spent catalysts were characterized to elucidate the deactivating mechanism. This presentation will discuss the reaction mechanism and the catalyst selection for this chemistry to produce the highest methyl benzaldehyde generation under various operating conditions.

A high selective catalyst for hydrogenation of 2-methylfuran to 2-methyltetrahydrofuran

Li Zeng-jie, Zhu Ming, Mei Hua*
College of Chemistry and Chemical Engineering,Nanjing University of Technology, Nanjing, China

Introduction

2-Methylfuran (2-MF) is a platform molecule that can produce many high valued chemicals by hydrogenation reaction [1,2]. 2-Methyltetrahydrofuran (2-MTHF) is one of hydrogenation products of 2-MF. 2-MTHF is widely used in agriculture, electronics and other industrial fields as a green solvent and an organic intermediate [3,4,5,6]. It is also used as biomass fuel because that over 60% of gasoline in motor engine can be replaced by 2-methyltetrahydrofuran without significant loss in engine performance [7]. 2-Methyltetrahydrofuran can be produced from glycols, inner esters and furfural. Considering that 2-methyl-1,4-butanediol used in glycol route is expensive and Co_{60} catalyst used in the inner esters route can usually cause environmental contamination[8], Furfural can be the promising raw material to produce 2-MTHF. The hydrogenation reaction involves two steps : First, 2-methylfuran is generated from furfural by -CHO hydrodeoxygenation over Cu-Cr catalysts [9,10]; Second, 2-methyltetrahydrofuran is synthesized by hydrogenation of 2-methylfuran over Pd or other non-noble metal catalysts. The first reaction has been wildly studied, thus we mainly focus on the second step of reaction-hydrogenation of 2-methylfuran to 2-methyltetrahydrofuran in this work (Figure 1).

Fig. 1 Reaction mechanism for hydrogenation of 2-methylfuran to 2-methyltetrahydrofuran

Although Pd catalyst has a high selectivity for hydrogenation of C=C bonds in a furan ring [11], the cost of Pd is expensive and limits its industrial application. Compared with noble metal, Ni based catalyst is less expensive and is easy to be prepared. Therefore, nickel has become a promising species to provide a high selectivity for 2-MF hydrogenation.

In this work, a series of hierarchical mesoporous Ni/Al_2O_3 catalyst were prepared by using impregnation method. Active alumina with large surface area was used as supportive material for Ni species. Catalytic performances for different Ni loading were investigated in a well-stirred batch reactor. Catalyst structure, SMSI (Support and Metal Strong Interactive) and redox behavior were measured by using N_2 adsorption-desorption (BET), X-ray Diffraction (XRD), H_2 temperature programmed reduction, etc. Then we evaluated catalytic performance of 2-MF hydrogenation reaction over several different catalysts.

Experimental Section

Preparation of Catalyst

Ni/Al_2O_3 catalyst was prepared by using impregnation method. Specifically, catalysts with different Ni contents were impregnated with an aqueous solution of the precursor salt. $Ni(NO_3)_2 \cdot 6H_2O$ was dissolved in deionized water to achieve Ni^{2+} mole concentration of 1mol/L.

Then activated alumina was added to the solution under electromagnetic stirring for 2h. The resulting mixture was aged overnight at room temperature and then it was evaporated at 80°C to remove water. Catalysts were dried up at 120 °C overnight, and they were calcinated in a furnace at 600°C for 4 h. Catalysts obtained using this method are designated as xNi/Al_2O_3 (where x denotes NiO mass fraction in the catalyst).

Catalyst Characterization

X-ray Diffraction (XRD) patterns were recorded on a Bruker AXS D8 advance diffractometer using Cu *Ka* radiation (30 mA, 40 kV). The scanning ranges were between 10° and 80° with rate of 20 deg s^{-1}.

BET surface areas were measured by N_2 adsorption-desorption isotherms at −196°C on a BELSORP Type II, Ankersmid company. Before measurements, samples were degassed under vacuum condition at 200°C for 3 hours.

H_2 temperature-programmed reduction (H_2-TPR) was carried out on a Quantachrome chemisorption in 10% of H_2 and 90% of Ar at a flow rate of 20 mL/min. Catalysts about 30 mg were degased in Ar at 200°C for 2h, followed by cooling to room temperature. Then reduction performance was conducted using temperature ramp of 10°C/min to 800°C.

Hydrogenation Test

The catalytic performances were evaluated using hydrogenation reaction of 2-MF to 2-MTHF. The hydrogenation reaction was carried out in a batch reactor with 15mL 2-MF and 0.5g catalyst. Before experiment, catalysts were reduced in flowing H_2 at 550°C for 4h, and air was evacuated in the system. Both temperature and pressure were increased to the setting points and were kept for four hours. The 2-MTHF products were determined with a gas chromatography system (GC-2014C) with capillary column (XE-60) and FID detector. Nitrogen was used as carrier gas for GC platform.

Results and Discussion

Catalyst Analysis

Table 1 is the BET analysis for activated alumina and γ-Al_2O_3. The specific surface area of activated alumina is 313 m^2/g, while surface area of γ-Al_2O_3 is 243 m^2/g. When activated alumina is calcinated at temperature of 450-600°C, crystal transforms to γ-Al_2O_3 and causes the decrease of surface area.

Table 2 shows surface area and pore structure of Ni/Al_2O_3 catalysts with various NiO loading. With NiO loading to 30%, the surface area of catalyst increases from 140 m^2/g to 152 m^2/g. The change of surface area can be attributed to pore-making effect when precursor materials decomposite and produce H_2O and NO_x during Ni/Al_2O_3 preparation [12,13]. When NiO content exceeds 30%, overloaded NiO may cover part of surface on supportive Al_2O_3, which leads to a decrease of pore volume and specific surface area.

Table 1 BET analysis for supports

Note: Activated alumina is used as support; γ-Al_2O_3 is prepared by calcination of activated alumina at 600°C.

Sample	*A* BET /($m^2 \cdot g^{-1}$)	Pore volume *v*/($cm^3 \cdot g^{-1}$)	Average pore diameter *d*/nm
activated alumina	313	0.4488	5.73
γ-Al_2O_3	243	0.4048	6.64

Table 2 BET analysis for calcined xNi/Al_2O_3 catalysts

NiO *w*/%	*A* BET /($m^2 \cdot g^{-1}$)	Pore volume *v*/($cm^3 \cdot g^{-1}$)	Average pore diameter *d*/nm
10	140	0.3242	8.64
25	149	0.3225	8.67
30	152	0.3486	9.17
40	108	0.2448	9.04

The XRD patterns of the prepared catalysts are presented in Figure 2. According to XRD patterns, activated alumina is a crystal mixture of χ,ρ,and γ-Al_2O_3. After calcination, the XRD pattern of catalyst shows the same signal with γ-Al_2O_3, which indicates that all crystals in activated alumina are changed into γ-Al_2O_3. In addition, characteristic diffraction peaks of Al_2O_3, $NiAl_2O_4$ and Ni are observed in Fig.2. The supportive material exhibits featured diffraction peaks of gamma alumina at 37.7°, 43.2° and 67.7°, while Ni species are observed at 44.5°, 51.9° and 76.4°. The $NiAl_2O_4$ characteristic diffraction peaks overlap γ-Al_2O_3 peaks [14]. Figure 2 shows a remarkable Ni diffraction peak when NiO content exceeds 25% in sample catalyst, which indicates that large Ni particles are formed. The XRD patterns in Fig.2 is consistent with BET results.

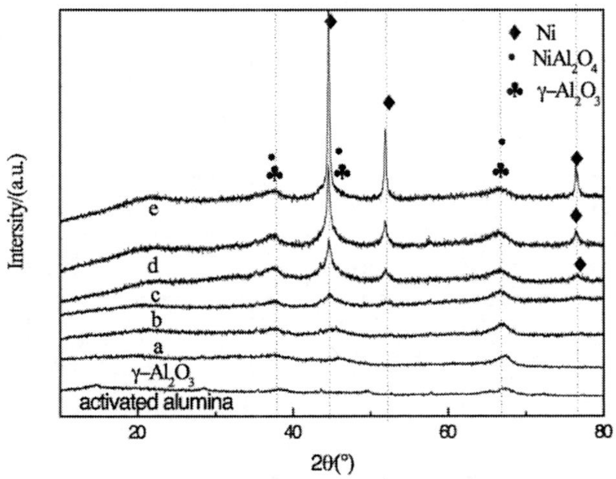

Fig.2 X-ray diffraction peaks of reduced Ni/Al_2O_3

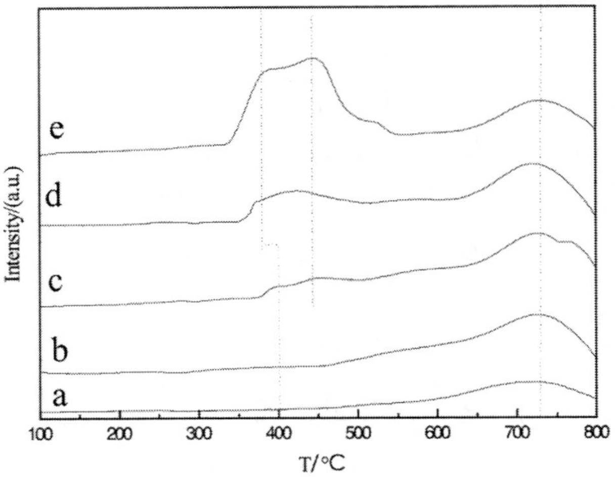

Fig.3 H_2-TPR spectra of Ni/Al_2O_3 catalysts
NiO loading: a-10%;b-20%;c-25%;d-30%;e-40%

Figure 3 shows the H_2-TPR characterization for the Ni/Al_2O_3 catalyst. For catalysts with Ni content between 25% and 40%, three peaks are detected in temperature range of 100-800°C. In Fig. 3, NiO reduction occurs at about 400 °C, which is in consistent with literature reports [15,16]. The low reduction temperature of NiO demonstrates a weaker interaction with supportive material. Compared with NiO, the high temperature peak (800°C) is assigned to the reduction of $NiAl_2O_4$, which indicates a strong interaction between Ni and supportive alumina. As shown in Fig.3, a small decrease of the reduction temperature is obtained with the increase of NiO loading, which accounts for the small SMSI during catalyst preparation.

Catalytic Performance

The effect of NiO content (10%, 20%, 25%, 30%, 35%, 40%) on catalytic performance of 2-methylfuran to 2-methyltetrahydrofuran at 150°C and 4MPa is summarized in Table 3. After 4 h of reaction in batch reactor, the product was cooled to room temperature and samples were analyzed by GC-2014C. Table 3 reports the conversion rate of 2-MF and the selectivity of hydrogenation reaction. As shown in Table 3, 80.2% of 2-MF conversion is obtained with NiO content of 10% on Ni/Al_2O_3 catalyst. However, 2-MF conversion reaches to 88.9% with NiO loading of 40%. In table 3, NiO content of 25% shows optimal catalytic performance for hydrogenation reaction.

Table 3 Effects of NiO contents on catalytic performance

NiO w/%	2-MF conversion x/%	Product selectivity s/%	
		2-MTHF	Others
10%	80.2	92.3	7.7
20%	98.9	94.3	5.7
25%	99.6	96.4	3.6
30%	99.3	93.6	6.3
35%	98.4	95.0	5.0
40%	88.9	95.4	4.6

Note: 2-MF: 15mL; Catalyst: 0.5g; Reaction pressure: 4MPa; Reaction temperature:150°C

According to BET isotherm, porous areas are increased with NiO content of 25%. The large surface area provides more active site for hydrogenation reaction. When NiO content exceeds 30%, NiO nano-particles are aggregated and inhibit pore diffusion inside 2-MF, and thus lead to the decreasing for catalytic performance. In Table 3, 2-MTHF selectivity achieves 96.4% and by-product reduces to 3.6% on 25% Ni/Al_2O_3 catalyst.

Table 4 Catalysts for 2-MF hydrogenation to 2-MTHF

Source	Catalyst	Pressure /MPa	Reaction temperature *t*/°C	2-MF conversion *x*/%	Product selectivity *s*/%	
					2-MTHF	Others
Fu [17]	Ni/SiO_2	1.6	180	98	86.8	13.2
Biswas [9]	Pd/C	1atm	220	100	<10%	-
Mo [18]	$Ru.Zn.Cu/SiO_2$	4	120	100	87.6	12.4
This work	Ni/Al_2O_3	3	150	99.4	97.1	2.9

We list several catalysts and their reaction performances for 2-MF hydrogenation to 2-MTHF in Table 4. Fu reported a Ni/SiO_2 catalyst could produce 2-MTHF via 2-MF hydrogenation at 1.6MPa and 180°C [17]. According to Mo et al. [18], 2-MTHF could be produced at relatively lower temperature of 120°C, but the operational pressure must increased to 4MPa. Biswas carried out the hydrogenation reaction at 1 atm and around 220°C, however its main production was 2-pentanone (2-PN) and the selectivity for 2-MTHF was less than 10% [9]. Although 2-MF conversion rates were close to 100%, 2-MTHF selectivity on most catalysts were below 90% according to literatures. At 3MPa and 150°C, the Ni/Al_2O_3 catalyst synthesized via impregnation method can achieve 97.1% for 2-MTHF selectivity.

The stability of synthesized catalyst-25%Ni/Al_2O_3 was analyzed at 3MPa, 150°C in a batch reactor with a mechanical stirring of 1000r/min. 0.04 g Fresh catalysts were added in the reactor to compensate weigh loss during each reaction. Experimental results are shown in Figure 4.

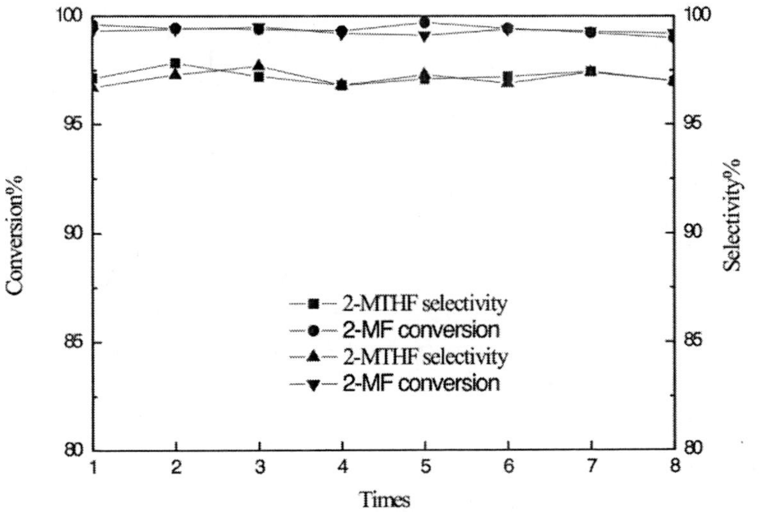

Fig. 4 Stability test for hydrogenation of 2-methylfuran to 2-methyltetrahydrofuran over 25%Ni/Al_2O_3

Note:●■ - fresh sample ; ▲▼ - used sample

In Fig. 4, 25%Ni/Al_2O_3 catalyst can be reused for at least 8 times without significant loss of conversion rate and product selectivity. The experimental results prove that the catalyst has good reproducibility and stability, and it could be used as a promising candidate for industrial practice.

Conclusions

In this study, active alumina with large specific surface area was used as supportive material. Ni/Al_2O_3 catalysts were synthesized by impregnation method. The experimental results show that NiO content on catalyst has an effect on the selectivity and conversion rate for hydrogenation of 2-methylfuran to 2-methyltetrahydrofuran. Catalytic performance is improved with NiO loading for the increase of active sites. However, overloaded NiO will inhibit hydrogenation reaction of 2-MF because mesoporous of supportive alumina are blocked. Therefore, Ni/Al_2O_3 with 25% NiO loading is considered to be a promising catalyst for industrial practice.

In batch reactor, the conversion rate of 2-methylfuran achieves 99.4% and the selectivity of 2-methyltetrahydrofuran reaches 97.1% on 25%Ni/Al_2O_3 catalyst at H_2 partial pressure of 3MPa, temperature of 150°C and with stirring speed of 1000r/min. The catalyst of Ni/Al_2O_3 can be reproducible for a large number of cycles.

References

1. Xiu S., Shahbazi A. (2012), "Bio-oil production and upgrading research: A review", *Renewable Sustain Energy Rev.*, pp. 4406-4414.
2. Zheng H., Zhu Y., Tend B., Ba Z., Zhang C., Xiang H., Li Y.(2006), "Towards understanding the reaction pathway in vapour phase hydrogenation of furfural to 2-methylfuran", *J. Mol. Catal. A: Chem.*, pp. 18-23.
3. Ragan J. A., Ende D. J., Brenek S. J., Eisnbeis S. A., Singer R. A., Tickner D. L., Teixeira J. J., Vanderplas B. C. and Westonet N. (2003), "Safe execution of a large-scale ozonolysis: preparation of the bisulfite adduct of 2-hydroxyindan-2-carboxaldehyde and its utility in a reductive amination", *Org. Process Res. Dev.*, pp. 155-160.
4. Yang J., Zheng H., Zhu Y., Zhao G., Zhang C., Teng B., Xiang H., Li Y. (2004), "Effects of calcination temperature on performance of Cu-Zn-Al catalyst for synthesizing c-butyrolactone and 2-methylfuran through the coupling of dehydrogenation and hydrogenation", *Catal. Commun.*, pp. 505-510.
5. Yan K., Wu G., LaFleur T., Jarvis C. (2014), "Production, properties and catalytic hydrogenation of furfural to fuel additives and value-added chemicals", *Renewable Sustain Energy Rev.*, pp. 663-676.
6. Nakagawa Y., Tamura M. and Tomishige K. (2013), "Catalytic reduction of biomass-derived furanic compounds with hydrogen", *ACS Catal.*, pp. 2655–2668.
7. Guo Q., Chen H. (2003), "Development on prepartion of levulinic acid and its derivatives", *Speciality Petrochemicals*, pp. 45-48.
8. Denney D. B., Denney D. Z. and Gigantinol J. J. (1984), "Cyclodehydration of 1,4-butanediols by pentaethoxyphosphorane", *J. Org. Chem.*, pp. 2831–2832.
9. Biswas P., Lin J. H., Kang J. and Guliants V. V. (2014), "Vapor phase hydrogenation of 2-methylfuran over noble and base metal catalysts", *Appl. Catal. A: General.*, pp. 379–385.

10. Sitthisa S., Sooknoi T., Ma Y., Balbuena P. B., Resasco D. E. (2011), "Kinetics and mechanism of hydrogenation of furfural on Cu/SiO_2 catalysts", *J. Catal.*, pp. 1-13.
11. Dong F., Zhu Y., Ding G., Cui J., Li X., Li Y. (2015), "One-step conversion of furfural into 2-methyltetrahydrofuran under mild conditions", *Chem. Sus. Chem.*, pp. 1534-1537.
12. Wang B., Shang Y., Ding G., Wang H., Wang E., Li Z., Ma X., Qin S., Sun Q. (2012), "Ceria-alumina composite support on the sulfur-resistant methanation activity of Mo-based catalyst", *J. Fuel Chem. Technol.*, pp. 1390-1396.
13. Sun J., Ren G., Huang Y., Chen X., Mei H. (2017), "Effect of calcination temperature on the catalytic performance of CuMgAl catalysts for furfural gas phase selective hydrogenation to furfuryl alcohol", *J. Fuel Chem. Technol.*, pp. 43-47.
14. Sepehri S., Rezaei M. (2015), "Preparation of highly active nickel catalysts supported on mesoporous nanocrystalline γ-Al_2O_3 for methane autothermal reforming", *Chem. Eng. Technol.*, pp. 1637-1645.
15. Bshish A., Yaakob Z., Ebshish A., Alhasan F. H. (2014), "Hydrogen production via ethanol steam reforming over Ni/Al_2O_3 catalysts: Effect of Ni loading", J. Energ. Resource-ASME, pp. 1-13.
16. Zhao A., Ying W., Zhang H., Ma H., Fang D. (2012), "Ni-Al_2O_3 catalysts prepared by solution combustion method for syngas methanation", *Catal. Commun.*, pp. 34-38.
17. Fu D., Yajing Z., Guijin Y., Kangjun W., Ileana D.,Valerian D., Yunfeng C., Jing W. (2015), "Synthesis and catalytic performance of Ni/SiO_2 for hydrogenation of 2-methylfuran to 2-methyltetrahydrofuran", *Journal of Nanomaterials*, pp. 60-62.
18. Mo Y., Wang G., Chen X., Feng N. (2013), "Preparation of 2-methyltetrahydrofuran by 2-methyfuran hydrogenation to Ru.Zn.Cu/SiO_2 composite catalyst", *Journal of Central South University of Forestry & Technology*, pp. 140-144.

538188 Glycerol-Free Biodiesel Production from Palm Oil and Supercritical Dimethyl/Diethyl Carbonate Mixtures with Trace Heterogeneous Base Catalyst

Abstract:

Palm oil, the representative component of waste fried oil, was used to produce glycerol-free biodiesel under supercritical dimethyl/diethyl carbonate (DMC/DEC) conditions at the catalysis of trace amount of heterogeneous potassium hydroxide (KOH). The single factor experimental design and Box-Behnken response surface methodology were combined to investigate the optimal reaction conditions and the interactive effects of variables. As a result, an optimal biodiesel yield of 91.3% (molar ratio of FAMEs/FAEEs=1.5:1) was obtained under milder reaction conditions: Reaction time is 34.7min, reaction temperature is 309.3°C, reaction pressure is 12MPa, molar ratio of DMC/DEC to oil is 26.9:1, molar ratio of DMC/DEC is 1:1 and catalyst amount of 0.2wt% (based on oil). Finally, the fuel properties of biodiesel were tested and compared with ASTM standard D-6751, and the cold flow performance of biodiesel had reached the requirements of commercial 0# diesel in China, with a cold filter plugging point of 3°C and a solidifying point of -2°C, respectively. Moreover, the differential scanning calorimetry (DSC) and rheological results showed that the low temperature property of biodiesel has been improved due to the excessive generation of ethyl esters during the supercritical method.

538798 Process Simulation and Analysis to Study Yield Conversion of Vegetable Oil to Biodiesel

Abstract:

Biodiesel is a sustainable biofuel due to its reduced emissions of sulfur oxides, nitrogen oxides and methane compared with derived fossil fuel. Biodiesel is commonly synthesized from the transesterification of vegetable oils using basic and acid homogeneous catalysts, a reaction which takes place in four steps to produce a mixture of fatty esters and glycerin. There are different works researching about using different precursors, catalysts, equipment as well as the conditions of the process to improve the biodiesel production. In this sense this paper reports the modeling of a reactor to carry out the study of the esterification and transesterification reactions that influence the biodiesel yield. This study was carried out using computational fluid dynamics where residence times, reactor geometry and feed rates are considered and it was found feed rates and reactor temperature increase considerable biodiesel yield.

538908 Modelling a Process to Study Production Biofuels

Abstract:

In this work was studied the process Fisher-Tropsch, which is used to convert the synthesis gas (H2 and CO) to obtain hydrocarbons from linear chains. To carry out this study, this process was divided into three stages, the first one corresponds to the production of synthesis gas obtained from a natural gas, water vapor and oxygen feed, using a Gibbs reactor under the operating conditions of 1066 ° C and 20 atm by reactions of methane reforming, partial oxidation and Dussan reaction using a molar ratio of 2.14 H2/CO. At second stage, production of synthetic fuels was carried out from synthesis gas considering that the reaction kinetics for the Fischer-Tropsch synthesis is governed by the carbene model, as well as by law of power of Langmuir-Hinshelwood. To simulate this stage, a furnace type multitubular packed bed reactor at a temperature of 232 ° C and 21 atm pressure were used, taking into account the selectivity of the alkanes, their frequency factors and the activation energy to obtain hydrocarbons of linear chains from methane to higher alkanes (triacontanes). Finally at third stage, synthetic fuels obtained were separated according to their octane and density through a rigorous distillation column given as results final products as: incondensable gases (C1-C4), gasoline (C5-C9), turbosine and kerosene (C10-C5), diesel (C15-C23) and heavy oils.

526660 Lipase-Catalyzed Synthesis of Biodiesel from Insect Fat Using Methyl Acetate As an Acyl Acceptor

Abstract:

Black soldier fly larvae (BSFL) are oleaginous insects that can assimilate organic waste for fat accumulation, and thus serve as an alternative feedstock for biodiesel production. In lipase-catalyzed transesterification, enzymes are deactivated by excess methanol. To address this obstacle, methyl acetate is suggested as an alternative acyl acceptor to methanol. In this study, methyl acetate was first used in the enzymatic production of biodiesel with BSFL as a triglyceride source. The interesterification of BSFL fat with methyl acetate was catalyzed using Novozym 435 as an efficient immobilized lipase. Response surface methodology was used to optimize the reaction and establish a reliable mathematical model for prediction. A maximum biodiesel yield of 96.97% was reached at a reaction time of 12 h, molar ratio of methyl acetate to fat of 14.64: 1, enzyme loading of 17.58%, and temperature of 39.5 °C. Under these optimal reaction conditions, Novozym 435 could be reused for up to 20 cycles without loss in enzyme activity. The properties of BSFL biodiesel were also investigated and all met the European standard EN 14214. This study indicates that the enzymatic interesterification of BSFL fat with methyl acetate is a promising and ecofriendly method for green fuel production

AUTHOR INDEX

AUTHOR INDEX

AUTHOR INDEX

AIChE
120 Wall Street, FL 23
New York, NY 10005-4020

ISBN 978-1-5108-7619-4

CURRAN ASSOCIATES INC.
proceedings.com